Agricultural Biotechnology, Forestry and Products

Agricultural Biotechnology, Forestry and Products

P. R. Gilbert

ANMOL PUBLICATIONS PVT. LTD.

NEW DELHI - 110 002 (INDIA)

ANMOL PUBLICATIONS PVT. LTD.
H.O.: 4374/4B, Ansari Road, Darya Ganj
New Delhi-110 002 (India)
Ph.: 23278000, 23261597
B.O.: 1015, Ist Main Road, BSK IIIrd Stage
IIIrd Phase, IIIrd Block
Bangalore - 560 085 (Karnataka)
Ph: 080-41723429, Tel/Fax: 080-2672 3604
Visit us at: www.anmolpublications.com

Agricultural Biotechnology, Forestry and Products

First Published, 2008

ISBN 978-81-261-3280-5

PRINTED IN INDIA

Printed at Mehra Offset Press, Delhi.

Contents

Preface

This book entitled "Agricultural Biotechnology, Forestry and Products" is designed to acquaint readers with the modern agricultural biotechnology, forestry and products at the global level. Agriculture of any type has an impact on the environment. Genetic engineering may accelerate the damaging effects of agriculture, have the same impact as conventional agriculture, or contribute to more sustainable practices. Growing genetically modified or conventional plants in the field has raised concern for the potential transfer of genes from cultivated species to their wild relatives. However, many food plants are not native to the areas in which they are grown. Locally, they may have no wild relatives to which genes could flow. Moreover, if gene flow occurs, it is unlikely that the hybrid plants would thrive in the wild, because they would have characteristics that are advantageous in agricultural environments only. In the future, genetically modified plants may be equipped with mechanisms designed to prevent gene flow to other plants. A controversy has arisen about whether certain genetically modified plants (which are insect resistant because they carry the Bt gene) could harm not only insect pests but also other species such as the monarch butterfly. In the field, no significant adverse effects on non-target species have so far been observed. Nonetheless, continued monitoring for such effects is needed. Genetically modified crops may have indirect environmental effects as a result of changing agricultural or environmental practices. However, it remains controversial whether the net effect of these changes will be positive or negative for the environment. For example, the use of genetically modified insect-resistant Bt crops is reducing the volume and frequency of insecticide use on maize, cotton and soybean. Yet the extensive use of herbicide and insect resistant crops could result in the emergence of resistant weeds and insects.

The broad consensus is that the environmental effects of genetically modified plants should be evaluated using science-based assessment procedures, considering each crop individually in comparison to its conventional counterparts. Animal feeds frequently contain genetically modified crops and enzymes derived from genetically modified micro-organisms. There is general agreement that both modified DNA and proteins are rapidly broken down in the digestive system. To date no negative effects on animals have been reported. It is extremely unlikely that genes may transfer from plants to disease-causing bacteria through the food chain. Nevertheless, scientists advise that genes which determine resistance to antibiotics that are critical for treating humans should not be used in genetically modified plants. As of 2004, no genetically modified animals were used in commercial agriculture anywhere in the world, but several livestock and aquatic species were being studied. Genetically modified animals could have positive environmental impacts, for example through greater disease resistance and lower antibiotic usage. However, some genetic modifications could lead to more intensive livestock production and thus increased pollution. Certain barriers to international agricultural trade have been reduced by the World Trade Organization (WTO). A WTO agreement adopted in 1994 establishes that countries retain their right to ensure that the food, animal, and plant products they import are safe. At the same time it states that countries should not use unnecessarily stringent measures as disguised barriers to trade. Several international agreements relate to the environmental aspects of genetically modified crops.

Agricultural biotechnology has international implications and may become increasingly important for developing countries. However, research has tended to focus on crops important to developed countries. To date, countries where genetically modified crops have been introduced in fields, have reported no significant health damage or environmental harm. Moreover, farmers are using less pesticides or using less toxic ones, reducing harm to water supplies and workers' health, and allowing the return of beneficial insects to the fields. Some of the concerns related to gene flow and pest resistance have been addressed by new techniques of genetic engineering. However, the lack of observed negative effects does not mean that they cannot occur. Scientists call for a cautious case-by-case assessment of each product or process prior to its release in order to address legitimate safety concerns. Science cannot declare any technology completely risk free. Genetically engineered crops can reduce some environmental risks associated with conventional agriculture, but will also introduce new challenges that must be addressed. Society will have to decide when and where genetic engineering is safe enough.

The discovery in 2001 of genetically engineered DNA in native Mexican corn varieties made concerns of genetic pollution actual, and led some scientists to worry that the spread of transgenes through cross-pollination could lead to a reduction in genetic diversity in important crops. Imports of genetically modified corn, soybeans, and other crops have been curtailed or limited in some countries, and the vast majority of such crops are grown in just a handful of nations. The Cartagena Protocol on Biosafety, which has been signed by more than 100 nations and took effect in Sept., 2003, requires detailed information on whether and how imported seeds, plants, animals, other organisms, and the like are genetically modified and permits a nation to bar those imports. The United States, however, is not party to the treaty. When Hungarian engineer Karl Ereky coined the word biotechnology in 1919 to describe the production of goods from raw materials with the aid of living organisms, he was describing techniques that had been employed by human beings for centuries. This includes techniques such as selective breeding, fermentation, hybridization, phyto-pharmacology, vaccination or the use of biomass for energy production. However, the term biotechnology has evolved to include modern techniques with new applications, to what is now known as modern biotechnology. Modern biotechnology encompasses techniques such as genetic engineering (which is the direct intervention on the genetic makeup of an organisms usually by introducing foreign DNA into its gene pool by means that would not occur naturally), tissue culture and others. These modern techniques also take considerably less time to achieve the desirable changes in living organisms and carry greater precision than traditional techniques. In short, genetic engineering is a specific subset as a part of the more general subject of biotechnology.

This book consists of—Agricultural Biotechnology: An Overview; Biotechnology, Farmers and Biotech Crops; Biotechnology in Forestry and Animal Agriculture; and Agricultural Biotech-Derived Products: A Case Study etc. besides a large body of list of acronyms, glossary of relevant terms, extensive bibliography, list of websites and links, for further referencing and research.

—Editor

List of Acronyms

AAAS	—	American Association for the Advancement of Science
AACC	—	American Association for Clinical Chemistry
AACR	—	American Association for Cancer Research
AAI	—	American Association of Immunologists
AANP	—	American Association of Neuropathologists
AAP	—	Association of American Physicians
ABA	—	Australian Biotechnol. Assoc.
ABI	—	Applied Biosystems, Inc.
ABIC	—	Agricultural Biotechnology Intl. Conf.
ABRF	—	Association of Biomolecular Resource Facilities
ACMG	—	American College of Medical Genet.
ACS	—	American Chemical Society
ADA	—	Americans with Disabilities Act
AEC	—	Atomic Energy Commission
AES	—	American Electrophoresis Society
AFMR	—	American Federation of Medical Research
AFIP/ARP	—	Armed Forces Inst. of Pathol./Am. Registry of Pathol.
AGSG	—	Alliance of Genetic Support Groups (now Genetic Alliance)
AGT	—	Association of Genetic Technologists

AHA — American Heart Association
AIBS — Am. Inst. of Biol. Soc.
AMIA — American Medical Informatics Association
ANGIS — Australian National Genomic Info. Service
ANL — Argonne National Laboratory
API — Application Programming Interface
ARP — American Registry of Pathology
ASB — American Society for Biotechnology
ASBMB — American Society for Biochemistry & Molecular Biology
ASCB — American Society for Cell Biology
ASCI — American Society for Clinical Investigation
ASHG — American Society for Human Genetics
ASIP — American Society for Investigative Pathologists
ASIS — American Society for Information Science
ASLME — American Society of Law, Medicine, and Ethics
ASM — American Society for Microbiology
ASPET — American Society for Pharmacology and Experimental Therapeutics
ATCC — American Type Culture Collection
ATP — Advanced Technology Program
AVS — American Vacuum Society
AWCH — Adelaide Women and Children's Hospital
BAC — Bacterial Artificial Chromosome
BACR — British Association for Cancer Research
BCM — Baylor College of Medicine
BDG — Batten's disease gene

BER	—	Biological and Environmental Research
BIO	—	Biotechnology Industry Organization
BNL	—	Brookhaven National Laboratory
BSCS	—	Biological Sciences Curriculum Study
BTCI	—	BioPharmaceutical Technology Center Institute
BTP	—	Biotechnology Training Programs
CAE	—	capillary array electrophoresis
CaSSS	—	California Separation Science Society
CATCMB	—	Center for Advanced Training in Cell and Molecular Biology
CCM	—	Chromosome Coordinating Meeting
CDC	—	Centers for Disease Control
cDNA	—	Complementary DNA
CE	—	Capillary Electrophoresis
CEPH	—	Centre d'Etude du Polymorphisme Humain
CF	—	Cystic Fibrosis
CFF	—	Cystic Fibrosis Foundation
CHH	—	Cartilage-hair Hypoplasia
CHI	—	Cambridge Healthtech Inst.
CHOP	—	Children's Hospital of Philadelphia
CIMB	—	Center for International Meeting on Biology
CIOMS	—	Council for International Organizations of Medical Sciences
CIRB	—	Colorado Institute for Research in Biotechnology
CLMA	—	Clinical Laboratory Management Association
cM	—	Centimorgan

CMT	—	Charcot-Marie-Tooth
CONTIG	—	Consortium of Teachers in Genetics
CORN	—	Council of Regional Networks for Genetic Services
CRADA	—	Cooperative Research and Development Agreement
CSHL	—	Cold Spring Harbor Laboratory
CTG	—	Trinucleotide Repeat
DHHS	—	Dept. of Health and Human Services
DIMACS	—	Center for Discrete Math. & Theoretical Comp. Sci.
DM	—	Myotonic Dystrophy
DMD	—	Duchenne Muscular Dystrophy
DNA	—	Deoxyribonucleic acid
DOE	—	Department of Energy
ECB	—	European Congress on Biotechnology
EDS	—	Electronic Data Submission
EEOC	—	Equal Employment Opportunity Commission
EL.B.A.	—	ELectronics and Biotechnology Advanced
EMBL	—	European Mol. Biol. Lab.
EMBO	—	European Molecular Biology Organisation
EMG	—	Encyclopedia of the Mouse Genome
EMS	—	Environmental Mutagen Society
EORTC	—	European Organization for Research and Treatment of Cancer
ERDA	—	Energy Research and Development Administration
ERI	—	Eleanor Roosevelt Institute
ES	—	Embryonic Stem

ESF	—	European Science Foundation
ESHG	—	European Society of Human Genetics
ESI	—	Electrospray Ionization
EST	—	Expressed Sequence Tag
EUCIB	—	European Collaborative Interspecific Backcross
EURESCO	—	European Research Conferences
FAP	—	Familial Adenomatous Polyposis
FASEB	—	Federation of American Societies for Experimental Biology
FCM	—	Flow Cytometry
FEBS	—	Fed. of Eur. Biochem. Soc.
FMF	—	Familial Mediterranean Fever
FRAXA	—	Fragile X locus
FSHD	—	Facioscapulohumeral Muscular Dystrophy
FTICR	—	Fourier Transform Ioncyclotron Resonance
FVEA	—	Fundacion Valenciana de Estudios Avanzados
GAS	—	Genome Automation System
GBASE	—	Genome Database of the Mouse
GBR	—	Global Business Research
GDB	—	Genome Database
GDB/OMIM	—	Genome Database/Online Mendelian Inheritance in Man
GESTEC	—	Genome Science and Technology Center
GIST	—	Genome Informatics System of Transputers
GLaRGG	—	Great Lakes Regional Genetics Group
GMCRF	—	General Motors Cancer Res. Foundation
GMD	—	Genomic Map Design

GPI	—	Genetics and Public Issues Program
GRAIL	—	Gene Recognition and Analysis Internet Link
GRC	—	Gordon Res. Conf.
GSA	—	Genetics Societies of America
GSDB	—	Genome Sequence Data Base
HAEC	—	Human Artificial Episomal Chromosome
HELSRD	—	Health Effects and Life Science Research Division
HERAC	—	Health and Environmental Research Advisory Committee
HGCC	—	Human Genome Coordinating Committee
HGM	—	Human Genome Meeting
HGMIS	—	Human Genome Management Information System
HGP	—	Human Genome Project
HHMI	—	Howard Hughes Medical Institute
HICSS	—	Hawaii Intl. Conf. on Systems Sci.
HLA	—	Human Leukocyte Antigen
HMDP	—	Homology Database
HNPCC	—	Hereditary Nonpolyposis Colorectal Cancer
HSA	—	Human Serum Albumin
HUGO	—	Human Genome Organisation
IARC	—	International Agency for Research on Cancer
IBC	—	International Business Communications
IBEX	—	International Biotechnology EXpo
IBF	—	International Business Forum
IBI	—	Institute for Biotechnology Information

ICES	—	International Council of Electrophoresis Societies
ICGEB	—	International Centre for Genetic Engineering and Biotechnology
ICHG	—	International Congress of Human Genetics
ICPEMC	—	International Commission on Protection Against Environmental Mutagens and Carcinogens
ICRF	—	Imperial Cancer Research Fund
IEEE	—	Institute of Electrical and Electronics Engineers
IG	—	IntelliGenetics
IGES	—	International Genetics Epidemiology Societies
IJCAI	—	International Joint Conference on Artificial Intelligence
IMA	—	Institute for Mathematics and its Applications
IMACS	—	International Association for Mathematics and Computers In Simulation
IMAGE	—	Integrated Molecular Analysis of Gene Expression
IMGS	—	Intl. Mammalian Genome Society
INRIA	—	French Natl. Inst. for Research in Computer Science and Control
IOM	—	Institute of Medicine
IOR	—	Institute of Religion
ISAG	—	Intl. Society for Animal Genetics
ISMB	—	Intelligent Systems for Molecular Biology
ISONG	—	Intl. Soc. of Nurses in Genet.

ISQL	—	Interactive Standard Query Language
ISTR	—	Institute for Science Training and Research
IUBMB	—	Intl. Union of Biochemistry and Molecular Biology
IU	—	Indiana University
IVD	—	*in vitro* Diagnostic
JGI	—	Joint Genome Institute
LANL	—	Los Alamos National Laboratory
LBNL	—	Lawrence Berkeley National Laboratory
LLNL	—	Lawrence Livermore National Laboratory
LTI	—	Life Technologies, Inc.
MALDI	—	matrix-assisted laser desorption ionization
MBC	—	Massachusetts Biotechnology Council
MBL	—	Marine Biological Laboratory
MCD	—	Mouse Cytogenetic Database
MDA	—	Muscular Dystrophy Assoc.
MEMS	—	Microelectromechanical Systems
MGC	—	Mouse Genome Conference
MGD	—	Mouse Genome Database
MGI	—	Microbial Genome Initiative
MHC	—	Major Histocompatibility Complex
MIMBD	—	Meet. on the Interconnection of Mol. Biol. Databases
MIT	—	Massachusetts Institute of Technology
MMRF	—	Marshfield Medical Research Foundation
MNBWS	—	Miami Nature Biotechnology Winter Symposium
MOD	—	March of Dimes
MOU	—	Memorandum of Understanding

MRC	—	Medical Research Council
MS	—	Mass Spectrometry
NACHGR	—	Natl. Advisory Council for Human Genome Res.
NAPBC	—	Natl. Action Plan on Breast Cancer
NAS	—	Natl. Academy of Sciences
NASA	—	Natl. Aeronautics and Space Administration
NBAC	—	Natl. Bioethics Advisory Commission
NCGR	—	National Center for Genome Resources
NCI	—	Natl. Cancer Institute
NCSA	—	National Center for Supercomputing Applications
NCSL	—	National Conference of State Legislatures
NCSU	—	North Carolina State University
NEH	—	National Endowment for the Humanities
NFCR	—	National Foundation for Cancer Research
NFID	—	National Foundation for Infectious Diseases
NHGRI	—	Natl. Human Genome Res. Inst.
NICHD	—	National Institute of Child Health and Human Development
NIGMS	—	National Institute of General Medical Sciences
NIH	—	National Institutes of Health
NIGMS	—	Natl. Institute of General Medical Sciences
NIST	—	Natl. Inst. of Standards and Technology
NLGLP	—	National Laboratory Gene Library Project
NLM	—	National Library of Medicine
NMHCC	—	National Managed Health Care Congress
NORD	—	National Organization for Rare Disorders

NRC	—	National Research Council
NRSA	—	National Research Service Award
NSF	—	Natl. Sci. Foundation
NSGC	—	National Society of Genetic Counselors
NSTA	—	National Science Teachers Association
NYAS	—	New York Academy of Science
OAINN	—	Ohio Aerospace Institute Neural Networks
OBER	—	Office of Biological and Environmental Research
OHER	—	Office of Health and Environmental Research
OMIM	—	Online *Mendelian Inheritance in Man*
OPRR	—	Office of Protection from Res. Risks
ORAU	—	Oak Ridge Associated Universities
ORISE	—	Oak Ridge Inst. for Science and Education
ORF	—	open reading frame
ORNL	—	Oak Ridge National Laboratory
OSU	—	Oregon State University
OTA	—	Office of Technology Assessment
PACHG	—	Program Advisory Committee on the Human Genome
PCR	—	Polymerase Chain Reaction
PFGE	—	Pulsed-field gel electrophresis
PG	—	Plant Genome
PIR	—	Protein Info. Resource
PNNL	—	Pacific Northwest National Laboratory
PRIM&R	—	Public Responsibility in Medicine & Research
PSC	—	Pittsburgh Supercomputing Center

QDFM	—	Quantitative DNA Fiber Mapping
RAPD	—	Random Amplified Polymorphic DNA
RARA	—	Retinoic Acid Receptor
RECOMB	—	Conference on Computational Molecular Biology
RFLP	—	Restriction Fragment Length Polymorphism
RH	—	Radiation Hybrid
RLGS	—	Restriction Landmark Genomic Scanning
SASE	—	Sample Sequencing
SBH	—	Sequencing by Hybridization
SBIR	—	Small Business Innovation Research
SCAN	—	Sequence Comparison ANalysis Program
SCE	—	School of Continuing Education
SCI	—	Society of Chemical Industry
SCW	—	Single-chromosome Workshop
SDC	—	San Diego Conference
SERGG	—	Southeast Regional Genetics Group
SHOM	—	Sequencing by Hybridization on Matrices
SIAM	—	Society for Industrial and Applied Mathematics
SIB	—	Society for Industrial Biology
SIM	—	Society for Industrial Microbiology
SIMS	—	Societal Institute of the Mathematical Sciences
SIVB	—	Society for In Vitro Biology
SNL	—	Sandia National Laboratory
SNP	—	Single Nucleotide Polymorphism
snRNP	—	Small Nuclear Ribonucleo Protein Particle

SQL	—	Standard Query Language
SSLP	—	Single-sequence Length Polymorphism
SSR	—	Simple Sequence Repeat
STM	—	Scanning Tunneling Microscope
STRP	—	Short Tandem Repeat Polymorphism
STS	—	Sequence-tagged Site
SULS	—	Suffolk Univ. Law School
TCR	—	T-cell receptor
TIGR/NIST	—	The Inst. for Genomic Res./Natl. Inst. of Standards and Technol.
UC	—	University of California
UCB	—	University of California, Berkeley
UCLA	—	University of California, Los Angeles
UICM	—	Univ. of Illinois College of Medicine
UMBC	—	University of Maryland, Baltimore County
UNC	—	University of North Carolina at Chapel Hill
UMDS	—	United Medical & Dental Schools (Univ. of London)
UNESCO	—	United Nations Educational, Scientific, and Cultural Organization
USDA	—	U.S. Dept. of Agriculture
UT	—	University of Tennessee
UW	—	University of Washington
WCCS	—	Workshop on Complete cDNA Sequencing
WHS	—	Wolf-Hirschhorn syndrome
WLMG	—	Wellcome Laboratory for Molecular Genetics

List of Glossary of Terms

Abiotic stress: Outside (nonliving) factors which can cause harmful effects to plants, such as soil conditions, drought, extreme temperatures.

Actin filament (microfilament): Helical protein filament formed by the polymerization of globular actin molecules. A major constituent of the cytoskeleton of all eucaryotic cells and part of the con- tractile apparatus of skeletal muscle.

Amino acid: Any of 20 basic building blocks of proteins—composed of a free amino (NH2) end, a free carboxyl (COOH) end, and a side group (R).

Amplify: To increase the number of copies of a DNA sequence, in vivo by inserting into a cloning vector that replicates within a host cell, or in vitro by polymerase chain reaction (PCR).

Anaerobe: An organism that grows in the absence of oxygen. See Aerobe.

Analyte: A chemical species targeted for qualitative or quantitative analysis.

Anneal: The pairing of complementary DNA or RNA sequences, via hydrogen bonding, to form a double-stranded polynucleotide. Most often used to describe the binding of a short primer or probe.

Antibiotic resistance: The ability of a microorganism to produce a protein that disables an antibiotic or prevents transport of the antibiotic into the cell.

Antibiotic: A class of natural and synthetic compounds that inhibit the growth of or kill other microorganisms.

Antibody: An immunoglobulin protein produced by B-lymphocytes of the immune system that binds to a specific antigen molecule.

Anticodon: A nucleotide base triplet in a transfer RNA molecule that pairs with a complementary base triplet, or codon, in a messenger RNA molecule. See Codon, Messenger RNA, RNA.

Antigen: Any foreign substance, such as a virus, bacterium, or protein, that elicits an immune response by stimulating the production of antibodies.

Antigenic determinant: A surface feature of a microorganism or macromolecule, such as a glycoprotein, that elicits an immune response.

Antigenic switching: The altering of a microorganism's surface antigens through genetic rearrangement, to elude detection by the host's immune system.

Antimicrobial agent: Any chemical or biological agent that harms the growth of microorganisms.

Anti-oncogene: See Recessive oncogene.

Antisense DNA: DNA strand that is complementary (opposite) to the functional gene. Antisense DNA can block the function of the normal sense DNA. This method was used in the construction of the Flavr Savr tomato.

Antisense RNA: A complementary RNA sequence that binds to a naturally occurring (sense) mRNA molecule, thus blocking its translation.

Aromatic: A term used to describe cyclic pi-bonded structures of special stability.

Asexual reproduction: Nonsexual means of reproduction which can include grafting and budding.

Autoproductivity: The ability of a system, under external control, to automatically produce an identical copy of itself.

Autosome: A chromosome that is not involved in sex determination.

Bacillus thuringiensis (Bt): A bacterium that kills insects; a major component of the microbial pesticide industry. A group of bacteria naturally occurring in soil that produce proteins, when ingested by specific insects, bind to their stomach lining (epithelium cell) receptors, and are toxic. There are thousands of strains of *Bacillus thuringiensis* each toxic to a limited group of insects such as beetles, flies, and moths. The advantage of crops engineered to produce Bt proteins is that they are toxic to a limited number of insects and are harmless to all mammals.

Backcross: Crossing an organism with one of its parent organisms.

Bacteriocide: A class of antibiotics that kills bacterial cells.

Bacteriophage (phage or phage particle): A virus that in- fects bacteria. Altered forms are used as vectors for cloning DNA.

Bacteriophage (phage): Any virus that infects bacteria. They were the first organisms used for the study of molecular genetics and are now widely used as cloning vectors. (From Greek *phagein*, to eat.)

Bacteriorhodopsin: Pigmented protein found in the plasma membrane of a salt-loving bacterium, *Halobacterium halobium*; it pumps protons out of the cell in response to light.

Bacteriostat: A class of antibiotics that prevents growth of bacterial cells.

Bacterium (plural bacteria): Common name for any member of the diverse group of procaryotic organisms. Most are single cells, but multicellular forms also exist.

Bacterium: A single-celled, microscopic prokaryotic organism: a single cell organism without a distinct nucleus.

Band: In gels and blots, bands are visible indications of a particular fragment of a certain size. There may be one to many bands per lane.

Banoprism: Scientists at Northwestern University have created a nanoparticle with a new shape that could be a useful tool in the race to detect biological threats. The nanoprism, which resembles a tiny Dorito, exhibits unusual optical properties that could be used to improve biodetectors, allowing them to test for a far greater number of biological warfare agents or diseases at one time.

Banoscience: The study of phenomena and manipulation of materials at atomic, molecular and macromolecular scales, where properties differ significantly from those at a larger scale. Nanoscience is primarily the extension of existing sciences into the realms of the extremely small (*nanomaterials, nanochemistry, nanobio, nanophysics*, etc.) while *nanoengineering* represents the extension of the engineering fields into the nano-scale realm (*nanofabrication, nanodevices*, etc.). The exponential growth of nanoscience is largely due to the development of new instruments and related techniques that are used to "routinely" probe and manipulate material at the atomic and molecular level. *Scanning probe microscopies,* analytical electron- beam techniques, epitaxial growth facilities, and *synchrotron* radiation sources are all opening huge opportunities.

Banoshells: Procedures that target cancer cells while leaving normal cells untouched, patients controlling the release of medicine in their bodies with an infrared light, and medical test results produced in seconds rather than days – these are three new medical technologies currently being tested by nanotechnology researchers at Rice University. ... The research focuses on nanoshells, a new type of nanoparticle invented by Naomi Halas,

Rice professor of electrical and computer engineering and chemistry. Nanoshells are layered nanoparticles whose ability to manipulate light and color can be designed into the nanoparticle by varying the thickness of the nanoparticle's layers.

Basal body: Short cylindrical array of microtubules plus their associated proteins found at the base of a eucaryotic cell cilium or flagellum. Serves as a nucleation site for the growth of the axoneme. Closely similar in structure to a centriole.

Basal lamina (plural basal laminae): Thin mat of extracellular matrix that separates epithelial sheets, and many types of cells such as muscle cells or fat cells, from connective tissue. Sometimes called a basement membrane.

Base pair (bp): A pair of complementary nitrogenous bases in a DNA molecule—adenine-thymine and guanine-cytosine. Also, the unit of measurement for DNA sequences. In DNA, there are four possible bases: cytosine (C), guanine (G), adenine (A), and thymine (T). Cytosine and thymine are pyrimidine bases; adenine and guanine are purine bases. Cytosine is complementary to guanine while adenine is complementary to thymine. If one strand of DNA has the sequence ATTGC then the complementary strand will be TAACG. Two complementary bases constitute a base pair. In RNA, thymine is replaced by uracil.

Beta-DNA: The normal form of DNA found in biological systems, which exists as a right-handed helix.

Beta-Lactamase: Ampicillin resistance gene.

Binding energy: The reduction in the free energy of a system that occurs when a ligand binds to a receptor. Generally used to describe the total energy required to remove something, or to take a system apart into its constituent particles—for example, to separate two atoms from one another, or to separate an atom into electrons and nuclei.

Bio-assemblies or Biomolecular Assemblies: Containing several protein units, DNA loops, lipids, various ligands, etc.

Bioaugmentation: Increasing the activity of bacteria that decompose pollutants; a technique used in bioremediation.

Biochauvinism: The prejudice that biological systems have an intrinsic superiority that will always give them a monopoly on self-reproduction and intelligence.

Biocompatible coated materials: Biocompatible materials usually used in dental and bone implants that enhance biologic fixation, thereby increasing the bond strength between the coated material and bone, and minimize possible biological effects that may result from the implant itself. MeSH, 1999

Biocompatible materials: Synthetic or natural materials, other than drugs, that are used to replace or repair any body tissue or bodily function. MeSH, 1973

Biodefense: The need for improved national defenses against biological attacks are urgently needed. A real-time reporting system needs to be developed, whereby officials can be informed about an emerging threat before an agent has a chance to affect thousands of people. The development of integrated systems for detecting and monitoring biological agents is instrumental to this goal. The development of vaccines and novel detection platforms will enable preparedness, while also benefiting human health.

Biodiversity: The wide diversity and interrelatedness of earth organisms based on genetic and environmental factors.

Bioengineering: Is rooted in physics, mathematics, chemistry, biology, and the life sciences. It is the application of a systematic, quantitative, and integrative way of thinking about and approaching the solutions of problems important to biology, medical research, clinical proactive, and population studies. The NIH Bioengineering Consortium agreed on the following definition for bioengineering research on biology, medicine, behavior, or health recognizing that no definition could completely eliminate overlap with other research disciplines or preclude variations in interpretation by different individuals and organizations. Integrates physical, chemical, or mathematical sciences and engineering principles for the study of biology, medicine, behavior, or health. It advances fundamental concepts, creates knowledge for the molecular to the organ systems levels, and develops innovative biologics, materials, processes, implants, devices, and informatics approaches for the prevention, diagnosis, and treatment of disease, for patient rehabilitation, and for improving health.

Bioenrichment: Adding nutrients or oxygen to increase microbial breakdown of pollutants.

Biofabrication: Includes nanoparticle delivery systems, biomaterials, tissue engineering, implants and prostheses. Using biological processes to synthesize and manufacture chemicals and materials of high value to the Department of Defense. Biological processes are characterized by: low energy barriers (~10Kcal, high temperatures/pressures not required); high reaction-, regio- and stereo- specificity (protection/ deprotection wastes and catalyst poisoning avoided); spatio-temporal control of materials synthesis of defined composition and size with angstrom- level precision; and local control of the dielectric environment (largely eliminating the need for toxic solvents). Potential target materials and chemicals include composites with enhanced mechanical properties, ultra-low k dielectrics and thermoelectrics, optoelectronic materials, photonic devices (waveguides and logic elements), electronic materials (e.g., GaN,

InGaN, AlGaN), elastomers, energetic materials, and adhesives. Biofabrication, DARPA.

Biologics: Agents, such as vaccines, that give immunity to diseases or harmful biotic stresses.

Biomass: The total dry weight of all organisms in a particular sample, population, or area.

Biomaterials: Synthetic or natural materials that can replace or augment tissues, organs or body functions.

Biomechanics: Mechanical structures of living organisms (especially muscles and bones).

Biomedical Nanotechnology: (1) the comprehensive monitoring, control, construction, repair, defense, and improvement of all human biological systems, working from the molecular level, using engineered nanodevices and nanostructures; (2) the science and technology of diagnosing, treating, and preventing disease and traumatic injury, of relieving pain, and of preserving and improving human health, using molecular tools and molecular knowledge of the human body; (3) the employment of molecular machine systems to address medical problems, using molecular knowledge to maintain and improve human health at the molecular scale. (4) A rapidly expanding field that includes many potential technologies and approaches. The key to this definition is that phenomena and materials at the nanometer scale are known to have properties that are uniquely attributable to that scale length. Nanomedicine could similarly be defined [as nanotechnology] as the design, synthesis, or application of materials, devices, or technologies in the nanometer-scale for the basic understanding, diagnosis, and / or treatment of disease. Canadian Institute of Health Research, Regenerative Medicine and Nanomedicine, RFA, 2003. The monitoring, repair, construction and control of human biological systems at the molecular level, using engineered nanodevices and nanostructures. (5) the comprehensive monitoring, control, construction, repair, defense, and improvement of all human biological systems, working from the molecular level, using engineered nanodevices and nanostructures; (6) the science and technology of diagnosing, treating, and preventing disease and traumatic injury, of relieving pain, and of preserving and improving human health, using molecular tools and molecular knowledge of the human body; (7) the employment of molecular machine systems to address medical problems, using molecular knowledge to maintain and improve human health at the molecular scale. The use of molecular-scale devices (nanotechnology) to repair damage and boost the immune system. Vapour grown carbon fibres are obtained as shown in this SEM image. The diameters of these fibres can vary from 100 nm to 500 nm. As part of the National Institutes of Health (NIH) Roadmap for Medical Research [nihroadmap.nih. gov], the NIH [nih.gov] will establish a handful of nanomedicine centers. These centers will be staffed by a highly interdisciplinary scientific crew including biologists, physicians, mathematicians, engineers and computer scientists. Research conducted over the first few years will be spent gathering extensive information about how molecular machines are built. A key activity during this time will be the development of a new kind of vocabulary, or lexicon, to define biological parts and processes in engineering terms. Once researchers have completely catalogued the interactions between and within molecules, they can begin to look for patterns and a higher order of connectedness than is possible to identify with current experimental methods. Mapping these networks and understanding how they change over time will be a crucial step toward helping scientists understand nature's rules of biological design.

Understanding these rules will, in many years' time, enable researchers to use this information to address biological issues in unhealthy cells.

Biomedical polymers: More and more therapeutic problems are relevant to the use of polymer- based therapeutic aids for a limited period of time, namely the healing time related to the outstanding capacity of living systems to self- repair ... After healing the remaining prosthetic materials or devices become foreign residues or wastes that have to be eliminated from the body. Nowadays, biocompatible polymers that can degrade in the body are developed. The degradation and the elimination of degradation by-products depend on rather complex phenomena that are presently reflected inconsistently by terms issued from the tradition because each domain has developed its own terminology almost independently. This is a source of misunderstandings, confusions and misperceptions among scientists, surgeons, pharmacists, journalists and politicians, the situation being increased by the introduction of degradable polymers in plastic waste management and environmental protection. Therefore, it is urgent to reflect the various phenomena by specific terms, harmonize and enforce their use by the people active in the biomedical, pharmacological and environmental fields, and, last but not least by the publishing media and journalists. IUPAC, Terminology for biomedical (therapeutic) polymers.

BioMEMS Biological MicroElectro Mechanical Systems: Highlights the technical advances in the field that are leading a revolution in medicine, and creating a new generation of analytical devices for medical diagnostics. The meeting will encompass technology developments in micro & nano drug delivery, interface of nanotech and tissue engineering, microfluidics, and miniaturized total analysis systems (microTAS), biosensors, innovations in mass spec, and nanoscale imaging

BioMEMS and Nanotech World, Aug. 16- 17, 2004, Washington DC

BioMEMS: MEMS used in medicine, that use microchips.

Biomimetic Chemistry: Knowledge of biochemistry, analytical chemistry, polymer science, and biomimetic chemistry is linked and applied to research in designing new molecules, molecular assemblies, and macromolecules having biomimetic functions. These new bio-related materials of high performance, including, for example, enzyme models, synthetic cell membranes, and biodegradable polymers, are prepared, tested, and constantly improved in this division for industrial scale production.

Biomimetic materials: Materials fabricated by BIOMIMETICS techniques, i.e., based on natural processes found in biological systems. [MeSH 2003] (2) Materials that imitate, copy, or learn from nature.

Biomimetic: Imitating, copying, or learning from nature. Nanotechnology already exists in nature; thus, nanoscientists have a wide variety of components and tricks already available. An interdisciplinary field in materials science, ENGINEERING, and BIOLOGY, studying the use of biological principles for synthesis or fabrication of BIOMIMETIC MATERIALS. There is a need to develop the next generation of restorative materials and medical implants. New avenues of scientific inquiry may enable the development of biomaterials that are safe, reliable, "smart", long-lasting, and perform ideally in their respective biological environments.... Over the last few years biomimetics and tissue engineering have emerged as a new vision in the field of tissue and organ repair and restoration. Biomimetics and tissue engineering are interdisciplinary fields that combine information from the study of biological structures and their functions with physics, mathematics, chemistry and engineering for the generation of new materials, tissues and organs. In the area of craniofacial, oral and dental principles from biomimetics and tissue engineering are applied to developing dental and facial implants, new polymers for guided tissue regeneration used in treating periodontal disease and bone and connective tissue defects, coral- based hydroxyapatite replicas for reconstruction of alveolar ridges and other osseous defects, temporomandibular joint (TMJ) and other joint prostheses, formation of bone matrix substitutes, and artificial replicas of bone, skin, and mucosa. The term biomimetics was coined in 1972 in the context of artificial enzymes; it might be defined broadly as "the abstraction of good design from nature". It is a fact of everyday life that nature has managed to built materials and 'devices' with breathtaking functionality, heterogeneity and stability by using a comparatively limited number of building blocks (the whole range of synthetic materials is restricted to man- made engineering). The basic concepts of nature are often simple; it is the way in which building blocks and materials are arranged that results in functionality. Among the most simple and abundant themes of nature is self- assembly: lipids assemble in sheets to form cell membranes, proteins assemble into functional enzymes, cellular 'sensors', fibers, or virus coats, and DNA assembles in double strands to provide the very basis for live: replication. Biomimetics Study of the structure and function of biological substances to make artificial products that mimic the natural ones.

BIOML Biopolymer Markup Language: Designed by the BIOML core team at Proteometrics, LLC and Proteometrics Canada Ltd. It is to be used "for the annotation of biopolymer sequence information. BIOML allows the full specification of all experimental information known about molecular entities composed of biopolymers, for example, proteins and genes."

Biomolecular materials: An emerging discipline, materials whose properties are abstracted from biology. They share many of the characteristics of biological materials but are not necessarily of biological origin. For example, they may be inorganic materials that are organized or processed in a biomimetic fashion. A key feature of biological and biomolecular materials is their ability to undergo *self- assembly*.

Biomolecular Nanotechnology: Nanotechnology existing in living systems and resulting from our ability to use biomolecules as components for molecular nanotechnology.

Biomotors: Driven by energy sources such as *adenosine triphosphate (ATP)* for chemical transduction and other processes. These biomotors are considered to be biomolecular and are discussed in the body of this report, but strictly speaking they do not conform to the panel's definition of self- assembly.

Bionanotechnology: Includes molecular motors, biomaterials, single molecule manipulation technologies, biochip technologies, etc.

Biopolymeroptoelectromechanical Systems [BioPOEMS]: Combining optics and microelectromechanical systems, and used in biological applications.

Biopolymers: Macromolecules (including proteins, nucleic acids and polysaccharides) formed by living organisms.

Bioremediation: The use of microorganisms to remedy environmental problems. See Bioaugmentation, Bioenrichment.

Biorobotics: Our research focuses on the role of sensing and mechanical design in motor control, in both robots and humans. This work draws upon diverse disciplines, including biomechanics, systems analysis, and neurophysiology. The main approach is experimental, although analysis and simulation play important parts. In conjunction with industrial partners, we are developing applications of this research in biomedical instrumentation, teleoperated robots, and intelligent sensors.

Biosensor: The term "biosensor" is a general designation that denotes either a sensor to detect a biological substance or a sensor which incorporates the use of biological molecules such as antibodies or enzymes. Biosensors are a subcategory of chemical sensors.

Biostasis: A condition in which an organism's cell and tissue structure are preserved, allowing later restoration by cell repair machines. Applicable to cryonics.

Biotechnology: The scientific manipulation of living organisms, especially at the molecular genetic level, to produce useful products. Gene splicing and use of recombinant DNA (rDNA) are major techniques used. Technology for working with biological systems. Includes genetic engineering, human and veterinary medicine, crop and animal breeding, diagnostics, pharmaceuticals, forensics, etc. Narrow sense: Genetic engineering.

Biotic stress: Living organisms which can harm plants , such as viruses, fungi, and bacteria, and harmful insects. See Abiotic stress.

Biovorous: From "biovore;" an organism capable of converting biological material into energy for sustenance.

Bipolar-junction transistor: Transistor with n-type and p-type semiconductors having base-emitter and collector-base junctions.

Blocks Substitution Matrix: A substitution matrix in which scores for each position are derived from *observations* of the frequencies of substitutions in blocks of local alignments in related proteins. Each matrix is tailored to a particular evolutionary distance. In the BLOSUM62 matrix, for example, the alignment from which scores were derived was created using sequences sharing no more than 62% identity. Sequences more identical than 62% are represented by a single sequence in the alignment so as to avoid over-weighting closely related family members. (Henikoff and Henikoff)

Blot, northern: A pattern of RNA fragments transferred to a nitrocellulose membrane from a gel. The gel has undergone electrophoresis to separate fragments according to size. The RNA fragments are arranged in different lanes for each sample. Each lane contains bands which are fragments of different sizes. Radioactive probes are often used to visualized particular bands by autoradiography.

Blot, southern: A pattern of DNA fragments transferred to a nitrocellulose membrane from a gel. The gel has undergone electrophoresis to separate fragments according to size. The DNA fragments are arranged in different lanes for each sample. Each lane contains bands which are fragments of different sizes. Radioactive probes are often used to visualized particular bands by autoradiography.

Blot, western: A pattern of proteins transferred to a nitrocellulose membrane from a gel. The gel has undergone electrophoresis to separate fragments according to size. The proteins are arranged in different lanes for each sample. Each lane contains bands which are fragments of different sizes. Labeled antibody probes are often used to visualize particular bands.

Bone substitutes: Synthetic or natural materials for the replacement of bones or bone tissue. They include hard tissue replacement polymers, natural coral, hydroxyapatite, beta- tricalcium phosphate, and various other biomaterials. The bone substitutes as inert materials can be incorporated into surrounding tissue or gradually replaced by original tissue.

Brownian Assembly: Brownian motion in a fluid brings molecules together in various position and orientations. If molecules have suitable complementary surfaces, they can bind, assembling to form a specific structure. Brownian assembly is a less paradoxical name for self-assembly (how can a structure assemble itself, or do anything, when it does not yet exist?).

Bt (*Bacillus thuringiensis*): A soil bacterium that produces insecticidal proteins. There are several different kinds of proteins produced by different strains of Bt. Some are effective against larvae of moths and butterflies. Others are effective against larvae of beetles. The Bt protein has been introduced into various crops as a built-in insecticide.

Calorie: Unit of heat. One calorie (small "c") is the amount of heat needed to raise the temperature of 1 gram of water by 1 °C. A kilocalorie is the unit used to describe the energy content of foods.

CAP (catabolite gene activator protein): Gene regulatory protein in procaryotes that, when glucose is absent, activates genes responsible for the breakdown of alternative carbon sources.

Capacitor: Energy storage circuit element having two conductors separated by an insulator.

Carcinogen: A substance that induces cancer.

Carcinoma: A malignant tumor derived from epithelial tissue, which forms the skin and outer cell layers of internal organs.

Catalyst: A substance that promotes a chemical reaction by lowering the activation energy of a chemical reaction, but which itself remains unaltered at the end of the reaction.

Catalytic antibody (abzyme): An antibody selected for its ability to catalyze a chemical reaction by binding to and stabilizing the transition state intermediate.

Catalytic RNA (ribozyme): A natural or synthetic RNA molecule that cuts an RNA substrate.

Cation: A positively charged ion.

cDNA library: A library composed of complementary copies of cellular mRNAs.

cDNA: DNA synthesized from an RNA template using reverse transcriptase. Complementary DNA to a particular RNA fragment.

Cell pharmacology: Delivery of drugs by medical nanomachines to exact locations in the body.

Cell Repair Machine: A system including nanocomputers and molecular scale sensors and tools, programmed to repair damage to cells and tissues. Molecular and nanoscale machines with sensors, nanocomputers and tools, programmed to detect and repair damage to cells and tissues, which could even report back to and receive instructions from a human doctor if needed.

Cell surgery: In medical nanorobotics, modifying cellular structures using medical nanomachines. Modifying cellular structures using medical nanomachines.

Cell: A membrane-bound unit, typically microns in diameter. All plants and animals are made up of one or more cells (trillions, in the case of human beings). In general, each cell of a multicellular organism contains a nucleus holding all of the genetic information of the organism. (2) A small structural unit, surrounded by a membrane, making up living things.

Cellular oncogene (proto-oncogene): A normal gene that when mutated or improperly expressed contributes to the development of cancer.

Centers of origin: Usually the location in the world where the oldest cultivation of a particular crop has been identified.

Central dogma: Francis Crick's seminal concept that in

nature genetic information generally flows from DNA to RNA to protein.

Centrifugation: Separating molecules by size or density using centrifugal forces generated by a spinning rotor. G forces of several hundred thousand times gravity are generated in ultracentrifugation.

Centromere: The central portion of the chromosome to which the spindle fibers attach during mitotic and meiotic division.

Chemotherapy: A treatment for cancers that involves administering chemicals toxic to malignant cells.

Chloramphenicol: An antibiotic that interferes with protein synthesis.

Chromatid: Each of the two daughter strands of a duplicated chromosome joined at the centromere during mitosis and meiosis.

Chromosome walking: Working from a flanking DNA marker, overlapping clones are successively identified that span a chromosomal region of interest.

Chromosome: A single DNA molecule, a tightly coiled strant of DNA, condensed into a compact structure in vivo by complexing with accessory histones or histone-like proteins. Chromosomes exist in pairs in higher eukaryotes. A linear structure in the nucleus of plants and animals that is visible in light microscopy when stained. The chromosome is a single, long, linear molecule of DNA and associated proteins. Bacteria have a single circular chromosome; other organisms may have many linear chromosomes.

Cistron: A DNA sequence that codes for a specific polypeptide; a gene. See DNA, Gene.

Clone: An exact genetic replica of a specific gene or an entire organism. See Cloning.

Cloning vector: A DNA molecule capable of autonomous

replication within the cloning host cell (e.g *E. coli*). The vectors contain restriction enzyme sites for insertion of foreign DNA. Cloning vectors are derived from bacterial plasmids, bacteriophages, or viruses.

Cloning: The mitotic division of a progenitor cell to give rise to a population of identical daughter cells or clones.

Coat protein (capsid): The coating of a protein that enclosed the nucleic acid core of a virus.

Codon: A group of three nucleotides that specifies addition of one of the 20 amino acids during translation of an mRNA into a polypeptide. Strings of codons form genes and strings of genes form chromosomes. Set of three nucleotides that specify a particular amino acid during protein synthesis.

Coenzyme (cofactor): An organic molecule, such as a vitamin, that binds to an enzyme and is required for its catalytic activity.

Colony: A group of identical cells (clones) derived from a single progenitor cell.

Combinatorial materials design: Uses computing power (sometimes together with massive parallel experimentation) to screen many different materials possibilities to optimize properties for specific applications (e.g., catalysts, drugs, optical materials).

Commensalism: The close association of two or more dissimilar organisms where the association is advantageous to one and doesn't affect the other(s). See Parasitism, Symbiosis.

Competency: An ephemeral state, induced by treatment with cold cations, during which bacterial cells are capable of uptaking foreign DNA.

Complementary DNA or RNA: The matching strand of a DNA or RNA molecule to which its bases pair.

Complementary nucleotides: Members of the pairs adeninethymine, adenine-uracil, and guaninecytosine that have the ability to hydrogen bond to one another.

Concatemer: A DNA segment composed of repeated sequences linked end to end.

Configuration space: A mathematical space describing the three-dimensional configuration of a system of particles (e.g., atoms in a nanomechanical structure) as a single point; the configuration space for an *N* particle system has 3*N* dimensions.

Conformation: The shape of a molecule. A molecular geometry that differs from other geometries chiefly by rotation about single or triple bonds; distinct conformations (termed con formers) are associated with distinct potential wells. Typical biomolecules and products of organic synthesis can interconvert among many conformations; typical diamondoid structures are locked into a single potential well, and thus lack conformational flexibility.

Conjugation: The joining of two bacteria cells when genetic material is transferred from one bacterium to another.

Constitutive promoter: An unregulated promoter that allows for continual transcription of its associated gene.

Contiguous (contig) map: The alignment of sequence data from large, adjacent regions of the genome to produce a continuous nucleotide sequence across a chromosomal region.

Coulomb [C]: Measure of electrical charge: 1 C is an amount of charge equal to that of about 6.24x1018 electrons.

Coupling: A connection between more than one object or energy pathway so that together they function as a single unit.

Cross-hybridization: The hydrogen bonding of a single-stranded DNA sequence that is partially but not entirely complementary to a singlestranded substrate. Often, this involves hybridizing a DNA probe for a specific

DNA sequence to the homologous sequences of different species.

Crossing-over: The exchange of DNA sequences between chromatids of homologous chromosomes during meiosis.

Cross-pollination: Fertilization of a plant from a plant with a different genetic makeup.

Crosstalk: Electromagnetic noise transmitted between leads or circuits in close proximity to each other.

Crybiology: The science of biology at low temperatures; research in cryobiology has made possible the freezing and storing of sperm and blood for later use.

Crystal Lattice: The regular three-dimensional pattern of atoms in a crystal.

Crystal structure: For crystalline materials, the manner in which atoms or ions are arrayed in space. It is defined in terms of the unit cell geometry and the atom positions within the unit cell.

Crystallescence: In medical nanorobotics, the crystallization of solid solute that is offloaded by nanorobot sorting rotors at a concentration that exceeds the solvation capacity of the surrounding solvent.

Culture: A particular kind of organism growing in a laboratory medium.

Curie temperature (also Curie point) (Tc): The temperature above which a ferromagnetic or ferrimagnetic material becomes paramagnetic. For iron the Curie point is 760oC and for nickel 356oC.

Current [A]: Measure of rate of flow of electric charge: a one-ampere current is a flow of 1 C of charge per second.

Cutoff: Condition in a diode or bipolar junction transistor in which the potential across a p-n junction prevents current flow.

Cyclic AMP (cyclic adenosine monophosphate): A second messenger that regulates many intracellular reactions by transducing signals from extracellular growth factors to cellular metabolic pathways.

Cyclic: A structure is termed cyclic if its covalent bonds form one or more rings.

Cytogenetics: Study that relates the appearance and behavior of chromosomes to genetic phenomenon.

Dalton: A unit of measurement equal to the mass of a hydrogen atom, 1.67 x 10E-24 gram/L (Avogadro's number).

Denature: To induce structural alterations that disrupt the biological activity of a molecule. Often refers to breaking hydrogen bonds between base pairs in double-stranded nucleic acid molecules to produce in single-stranded polynucleotides or altering the secondary and tertiary structure of a protein, destroying its activity.

Density gradient centrifugation: High-speed centrifugation in which molecules "float" at a point where their density equals that in a gradient of cesium chloride or sucrose.

Diabetes: A disease associated with the absence or reduced levels of insulin, a hormone essential for the transport of glucose to cells.

Dideoxynucleotide (didN): A deoxynucleotide that lacks a 3' hydroxyl group, and is thus unable to form a 3'-5' phosphodiester bond necessary for chain elongation. Dideoxynucleotides are used in DNA sequencing and the treatment of viral diseases.

Digest: To cut DNA molecules with one or more restriction endonucleases.

Diploid cell: A cell which contains two copies of each chromosome. See Haploid cell.

Directional cloning: DNA insert and vector molecules are digested with two different restriction enzymes to create noncomplementary sticky ends at either end of each restriction fragment. This allows the insert to be ligated to the vector in a specific orientation and prevents the vector from recircularizing.

DNA: DNA is the recipe for life. DNA is a molecule found in the nucleus of every cell and is made up of 4 nucleotides Adenine (A), Guanine (G), Thymine (T), and Cystosine (C). The order of the nucleotides in the DNA strand holds a code of information for the cell.

DNA (Deoxyribonucleic acid): An organic acid and polymer composed of four nitrogenous bases—adenine, thymine, cytosine, and guanine linked via intervening units of phosphate and the pentose sugar deoxyribose. DNA is the genetic material of most organisms and usually exists as a double-stranded molecule in which two antiparallel strands are held together by hydrogen bonds between adeninethymine and cytosine-guanine.

DNA (Deoxyriobonucleic Acid): DNA molecules are long chains consisting of four kinds of nucleotides; the order of these nucleotides encodes the information needed to construct protein molecules. These in turn make up much of the molecular machinery of the cell. DNA is the genetic material of cells.

DNA Chip: Also: Gene Chip and DNA Microchip. A purpose built microchip used to identify mutations or alterations in a gene's DNA.

DNA diagnosis: The use of DNA polymorphisms to detect the presence of a disease gene.

DNA diagnostics - miniaturization of: In the areas of sample preparation and assay, it is clear that miniaturization is key. To reduce the size of samples by a factor of 10 or greater, barriers in microfluidics, micromachining, robotics, microchemistry, nucleic acid chemistry, and surface chemistry must be overcome. To implement miniaturized protocols accurately and efficiently, substantial automation of the process will be required. In the development of miniaturized systems, it is essential that the system can be adapted for high levels of parallelization. Miniaturization poses significant technological risks. Currently, there exists no universally accepted precedent for the handling, replication, amplification, or cloning of DNA in nanoliter volumes. Due to the size and charge of the DNA molecule, and the relative instability of many of the enzymes involved in the sample preparation processes, nanoliter and less volumes may pose substantial challenges. In addition, interactions of the biological molecules with the surfaces of the reaction chambers must be minimized. For some methodologies, it is not clear what the optimal sample will be, so substantial improvement in DNA fragmentation technologies or DNA cloning vectors may be required for the ultimate efficient application to diagnostics. Improvements in any of these areas are likely to be of value to other non- DNA based diagnostic applications such as antibody screening protocols and enzyme based diagnostics, because miniaturized robotic or micro- electro mechanical systems developed for DNA could be modified to be used for these purposes.

DNA fingerprint: The unique pattern of DNA fragments identified by Southern hybridization (using a probe that binds to a polymorphic region of DNA) or by polymerase chain reaction (using primers flanking the polymorphic region).

DNA polymorphism: One of two or more alternate forms (alleles) of a chromosomal locus that differ in nucleotide sequence or have variable numbers of repeated nucleotide units.

DNA Probes: A DNA or nucleic acid probe is a short strand of DNA that locates and binds to its complementary sequence in samples containing single strands of DNA or RNA enabling identification of specific sequences. Nucleic acid probe assays exploit the fundamental hybridization reaction that occurs spontaneously between two complementary DNA:DNA or DNA:RNA strands. As in immunoassays, detection of the hybrid requires that the probe be labeled. Various direct and indirect methods have been devised for the detection of the hybrid. Direct labeling involves attaching the label directly to the probe sequence; indirect labeling binds an antibody to the DNA:DNA or DNA:RNA hybrid. As in immunoassays, non-isotopically-labeled probes are preferred over radio-labeled probes primarily because of radiation hazards, disposal problems, and short reagent shelf life. In addition, the factors determining the detection limits of hybridization assays based on labeled probes are similar to those in immunoassays. Therefore, the development of a simple, inexpensive and sensitive direct detection system which eliminates the use of labels is highly desirable.

DNA Sequencing: There are two main classical methods for sequencing DNA: The first method, developed by Allan Maxam and Walter Gilbert, involves chemicals used to cleave the DNA at certain positions, generating a set of fragments that differ by one nucleotide. The second method, developed by Fred Sanger and Alan Coulson, involves enzymatic synthesis of DNA strands that terminate in a modified nucleotide. Analysis of fragments is similar for both methods and involves gel electrophoresis and autoradiography or fluorescence. The enzymatic method has largely replaced the chemical method as the technique of choice, although there are some situations where chemical sequencing can provide data more easily than the enzymatic method.

DNA, recombinant: DNA that has been cut and spliced back together in a new sequence. The DNA may be from one organism or from more than one organism

DNA, repetitive: Fragments of DNA that appear in multiple copies in a single individual.

DNA: A molecule encoding genetic information, found in the cell's nucleus. Deoxyribose Nucleic. Acid. DNA is a molecule used in biological systems to store genetic information. It has a long, double stranded structure with the two strands wrapped around each other to form a right handed helix. There are about 10 nucleotide pairs per helical turn. Each strand is composed of a sugar phosphate backbone and attached bases. It is connected to a complementary strand by hydrogen bonding (non- covalent) between paired bases, adenine (A) with thymine (T) and guanine (G) with cytosine (C). DNA is an abbreviation of DeoxyriboNucleic Acid. : It is the genetic 'blueprint' of all life on this planet. : DNA typing is often applied to Human Identification which has also been called DNA Fingerprinting. DNA Fingerprinting has been attributed to Alex Jeffries, who with his colleagues in 1985, discovered a sequence of nucleotides, subsequently termed short tandem repeats (STR) when comparing the DNA sequence of the human á-globin chromosomes isolated from different individuals.

DNA: Deoxyribonucleic acid, the molecule that stores genetic information. Composed of two complementary strands. See base pairs.

Dominant gene: A gene whose phenotype is when it is present in a single copy.

Dominant(-acting) oncogene: A gene that stimulates cell proliferation and contributes to oncogenesis when present in a single copy.

Dominant: An allele is said to be dominant if it expresses its phenotype even in the presence of a recessive allele. See Allele, Phenotype, Recessive.

Dormancy: A period in which a plant does not grow, awaiting necessary environmental conditions such as temperature, moisture, nutrient availability.

Double helix: Describes the coiling of the antiparallel strands of the DNA molecule, resembling a spiral staircase in which the paired bases form the steps and the sugar-phosphate backbones form the rails.

Double-stranded complementary DNA (dscDNA): A duplex DNA molecule copied from a cDNA template.

Downstream: The region extending in a 3' direction from a gene.

Ecology: The study of the interactions of organisms with their environment and with each other.

Ecosystem: The organisms in a plant population and the biotic and abiotic factors which impact on them. See abiotic factors; Biotic factors.

Electrophoresis: The technique of separating charged molecules in a matrix to which is applied an electrical field.

Electroporation: A method for transforrning DNA, especially useful for plant cells, in which high voltage pulses of electricity are used to open pores in cell membranes, through which foreign DNA can pass.

Electroporation: Transformation technique that uses electric fields to temporarily increase permeability of cells to foreign DNA.

Emergence: A complex whole created by simple parts, as in the brain where billions of neurons work individually, but collectively make up our consciousness and give us the ability to think, rationalize, and create.

Emulation: An absolutely precise simulation of something, so exact that it is equivalent to the original (for example, many computers emulate obsolete computers to run their programs). The Star Trek replicator is an example.

Enabling science and technologies: Areas of research relevant to a particular goal, such as nanotechnology. Also, technology that "enables" other technology to advance, such as the transistor enabled the computer chip revolution, as did photolithography.

Encapsidation: Process by which a virus' nucleic acid is enclosed in a capsid. See Coat protein.

Endocrine cell: Specialized animal cell that secretes a hormone into the blood; usually part of a gland, such as the thyroid or pituitary gland.

Endocytosis: Uptake of material into a cell by an invagination of the plasma membrane and its internalization in a membrane-bounded vesicle.

Endophyte: An organism that lives inside another.

Endoplasmic reticulum (ER): Labyrinthine, membrane-bounded compartment in the cytoplasm of eucaryotic cells, where lipids are synthesized and membrane-bound proteins are made.

Endosome: Membrane-bounded organelle in animal cells that carries materials newly ingested by endocytosis and passes many of them on to lysosomes for degradation.

Endothelium: Single sheet of highly flattened cells (endothelial cells) that forms the lining of all blood vessels. Regulates exchanges between the bloodstream and surrounding tissues and is usually surrounded by a basal lamina.

Energy [J]: Capacity for performing work or to cause heat flow. Like work itself, it is measured in Joules.

Environmental Protection Agency (EPA): The U.S. regulatory agency for biotechnology of microbes. The major laws under which the agency has regulatory powers are the Federal Insecticide, Fungicide, and Rodenticide Act (FIFRA); and the Toxic Substances Control Act (TSCA).

Enzymes: Proteins that control the various steps in all chemical reactions.

EPROM: Electrically Programmable Read-Only Memory — nonvolatile memory device.

Escherichia coli: A commensal bacterium inhabiting the human colon that is widely used in biology, both as a simple model of cell biochemical function and as a host for molecular cloning experiments.

Ethidium bromide: A fluorescent dye used to stain DNA and RNA. The dye fluoresces when exposed to UV light.

Eukaryote: An organism whose cells possess a nucleus and other membrane-bound vesicles, including all members of the protist, fungi, plant and animal kingdoms; and excluding viruses, bacteria, and blue-green algae. See Prokaryote.

Eurisko: A computer program developed by Professor Douglas Lenat which is able to apply heuristic rules for performing various tasks, including the invention of new heuristic rules.

Event: An "event" in genetic engineering is the insertion of a particular piece of foreign DNA into the chromosome of the recipient. Insertion occurs in random locations, so each event is unique. The event can affect how a gene is expressed in the organism. Once an event occurs, the transgene can be passed to the next generation as a normally inherited gene.

Evolution: The long-term process through which a population of organisms accumulats genetic changes that enable its members to successfully adapt to environmental conditions and to better exploit food resources.

Exon: A DNA sequence that is ultimately translated into protein. See DNA.

Expectation value: The number of different alignents with scores equivalent to or better than S that are expected to occur in a database search by chance. The lower the E value, the more significant the score.

Express: To translate a gene's message into a molecular product.

Face-centered cubic: A crystal structure found in common elemental metals also FCC. Within the cubic unit cell, atoms are located at all corner and face-centered positions.

Fail-stop: Describes a component or subsystem that, in the event of a failure, produces no output (e.g., of material or data) rather than producing a damaged or incorrect output.

Fermi energy: The energy level in a solid at which the probability of finding an electron is 1/2. For a metal, the energy corresponding to the highest filled electron state in the valence band at 0 K.

Field-Effect Transistor (FET): Field-Effect Transistor — semiconductor device whose insulated gate electrode controls current flow.

FIFRA: The Federal Insecticide, Fungicide, and Rodenticide Act. See Environmental Protection Agency.

Fingerprint: A set of molecular markers sufficiently diverse to identify particular individuals with reasonable certainty.

Flanking region: The DNA sequences extending on either side of a specific locus or gene.

Food and Drug Administration (FDA): The U.S. agency responsible for regulation of biotechnology food products. The major laws under which the agency has regulatory powers include the Food, Drug, and Cosmetic Act; and the Public Health Service Act.

Fusion gene: A hybrid gene created by joining portions of two different genes (to produce a new protein) or by joining a gene to a different promoter (to alter or regulate gene transcription).

Gamete: A haploid sex cell, egg or sperm, that contains a single copy of each chromosome.

Ganglioside: Any glycolipid having one or more sialic acid residues in its structure. Found in the plasma membrane of eucaryotic cells and especially abundant in nerve cells.

Gel, agarose: A substance used to separate DNA or RNA fragments by size. Used for southern and northern blots.

Gel, polyacrylamide: A substance used to separate proteins by size. Used for western blots.

GEM: A genetically engineered microorganism.

Gene amplification: The presence of multiple genes. Amplification is one mechanism through which proto-oncogenes are activated in malignant cells.

Gene cloning: The process of synthesizing multiple copies of a particular DNA sequence using a bacteria cell or another organism as a host. See DNA, Host.

Gene expression: The process of producing a protein from its DNA- and mRNA-coding sequences.

Gene flow: The exchange of genes between different but (usually) related populations.

Gene frequency: The percentage of a given allele in a population of organisms. See Allele.

Gene gun: Transformation technique that uses accelerated particles coated with DNA to introduce foreign DNA into recipient plant.

Gene insertion: The addition of one or more copies of a normal gene into a defective chromosome.

Gene linkage: The hereditary association of genes located on the same chromosome.

Gene modification: The chemical repair of a gene's defective DNA sequence. See DNA.

Gene pool: The totality of all alleles of all genes of all individuals in a particular population.

Gene therapy: The introduction of new genes into individuals to cure diseases or genetic abnormalities.

Gene translocation: The movement of a gene fragment from one chromosomal location to another, which often alters or abolishes expression.

Gene: The basic unit of inheritance. A segment of DNA that codes for a particular protein. A locus on a chromosome that encodes a specific protein or several related proteins. It is considered the functional unit of heredity.

Genetic assimilation: Eventual extinction of a natural species as massive pollen flow occurs from another related species and the older crop becomes more like the new crop. See Gene flow.

Genetic code: The three-letter code that translates nucleic acid sequence into protein sequence. The relationships between the nucleotide base-pair triplets of a messenger RNA molecule and the 20 amino acids that are the building blocks of proteins. See Base pair, Nucleic acid, Nucleotide. The genetic information in DNA is encoded with four different nucleotide bases: A, C, G, and T. A set of three consecutive nucleotide bases constitutes a codon. A codon specifies a particular amino acid that is added during synthesis of a protein.

Genetic disease: A disease that has its origin in changes to the genetic material, DNA. Usually refers to diseases that are inherited in a Mendelian fashion, although noninherited forms of cancer also result from DNA mutation.

Genetic drift: Random variation in gene frequency from one generation to another.

Genetic engineering: Genetic engineering is the process of manually adding new DNA on a molecular level with the goal of adding one or more new traits that are not already found in that organism. For the purposes of this web site, genetic engineering includes recombinant DNA and gene splicing technologies.

Genetic linkage map: A linear map of the relative positions of genes along a chromosome. Distances are established by linkage analysis, which determines the frequency at which two gene loci become separated during chromosomal recombination.

Genetic marker: A gene or group of genes used to "mark" or track the action of microbes.

Genetic modification: The selective and purposeful alteration of genes by humans. Genetic modification includes methods of incorporating new genes or traits into an organism through genetic engineering, mutagenesis and traditional breeding methods.

Genome: The genetic complement contained in the chromosomes of a given organism, usually the haploid chromosome state. The full chromosome set containing all the genes of a particular individual.

Genomic library: A library composed of fragments of genomic DNA.

Genomics: The study of the structure and function of genomes. Genomics usually involves high speed sequencing of the DNA and computer searches for sequences that code for genes.

Genotype: The sum total of the genetic information of an organism including the linkage relationships between genes. The genotype, modified by environment, determines the phenotype.

Genus: A category including closely related species. Interbreeding between organisms within the same category can occur.

Germ cell (germ line) gene therapy: The repair or replacement of a defective gene within the gamete-forming tissues, which produces a heritable change in an organism's genetic constitution.

GMO: Genetically modified organism. An organism that has incorporated a functional foreign gene through recombinant DNA technology. The novel gene exists in all of its cells and is passed through to progeny. Same as transgenic.

Green revolution: Advances in genetics, petrochemicals, and machinery that culminated in a dramatic increase in crop productivity during the third quarter of the 20th century.

Growth curve: See Growth phase.

Growth factor: A serum protein that stimulates cell division when it binds to its cell-surface receptor.

Growth phase (curve): The characteristic periods in the growth of a bacterial culture, as indicated by the shape of a graph of viable cell number versus time.

GUI: Graphical User Interface — hardware, software, and firmware that produces the display on modern personal computers.

Guy Fawkes Scenario: If nanotechnology becomes widely available, it might become trivial for anyone to committ acts of terrorism (such as making nanomachines build a large amount of explosives under government buildings a la Guy Fawkes). This would either force strict control over nanotechnology (hard) or a decentralized mode of organization.

Hair cell: Specialized sensory epithelial cell in the ear with bundles of giant microvilli (stereocilia) protruding from its apical surface. Sound vibrations tilt the stereocilia, evoking an electrical change in the hair cell, which thus acts as a sound detector.

Hall effect: The phenomenon whereby a force is applied to a moving electron or hole by a magnetic field that is applied perpendicular to the direction of motion. The force direction is perpendicular to both the magnetic field and the particle motion directions.

Haploid cell: A cell containing only one set, or half the usual (diploid) number, of chromosomes.

Heme: An iron complex. Cyclic organic molecule containing an iron atom that carries oxygen in hemoglobin and carries an electron in cytochromes.

Hemophilia: An X-linked recessive genetic disease, caused by a mutation in the gene for clotting factor VIII (hemophilia A) or clotting factor IX (hemophilia B), which leads to abnormal blood clotting.

Herbicide resistance: Herbicide resistance is when plants become resistant to the negative effects of a particular herbicide formulation or are genetically engineered in order to have resistance to a particular herbicide. Herbicide-resistant crops can be sprayed with a particular herbicide, killing weeds growing in and around the crop plants, without being damaged.

Herbicide: Any substance that is toxic to plants; usually used to kill specific unwanted plants.

Heterochromatin: Dark-stained regions of chromosomes thought to be for the most part genetically inactive.

Heteroduplex: A double-stranded DNA molecule or DNA-RNA hybrid, where each strand is of a different origin.

Heterogeneous nuclear RNA (hnRNA): The name originally given to large RNA molecules found in the nucleus, which are now known to be unedited mRNA transcripts, or pre-mRNAs.

Homologous chromosomes: Chromosomes that have the same linear arrangement of genes—a pair of matching chromosomes in a diploid organism. See Chromosomes.

Homologous recombination: The exchange of DNA fragments between two DNA molecules or chromatids of paired chromosomes (during crossing over) at the site of identical nucleotide sequences.

Homozygote: An organism whose genotype is characterized by two identical alleles of a gene. See Allele, Genotype.

Host: An organism that contains another organism.

Human Genome Project: A project coordinated by the National Institutes of Health (NIH) and the Department of Energy (DOE) to determine the entire nucleotide sequence of the human chromosomes.

Human growth hormone (HGH, somatotrophin): A protein produced in the pituitary gland that stimulates the liver to produce somatomedins, which stimulate growth of bone and muscle.

Hybrid: The offspring of two parents differing in at least one genetic characteristic (trait). Also, a heteroduplex DNA or DNA-RNA molecule.

Hybridization, nucleic acid: Complementary strands of DNA or RNA will spontaneously match up with each other and bond together under the right conditions. This is called nucleic acid hybridization and is used when probing southern and northern blots.

Hybridization: The hydrogen bonding of complementary DNA and/or RNA sequences to form a duplex molecule.

Hybridoma: A hybrid cell, composed of a B Iymphocyte fused to a tumor cell, which grows indefinitely in tissue culture and is selected for the secretion of a specific antibody of interest.

Hydrogen bond: A relatively weak bond formed between a hydrogen atom (which is covalently bound to a nitrogen or oxygen atom) and a nitrogen or oxygen with an unshared electron pair.

Hydrolysis: A reaction in which a molecule of water is added at the site of cleavage of a molecule to two products.

Imaging Contrast Agent: A molecule or molecular complex that increases the intensity of the signal detected by an imaging technique, including MRI and ultrasound. An MRI contrast agent, for example, might contain gadolinium attached to a targeting antibody. The antibody would bind to a specific target—a metastatic melanoma cell, for example – while the gadolinium would increase the magnetic signal detected by the MRI scanner.

Immortalization: Production of a cell line capable of an unlimited number of cell divisions. Can be the result of a chemical or viral transformation or of fusion with cells of a tumor line.

Immortalizing oncogene: A gene that upon transfection enables a primary cell to grow indefinitely in culture.

In situ: Refers to performing assays or manipulations with intact tissues.

In vivo: Refers to biological processes that take place within a living organism or cell.

Incomplete dominance: A condition where a heterozygous off- spring has a phenotype that is distinctly different from, and intermediate to, the parental phenotypes. See Heterozygote, Phenotype.

Initiation codon: The mRNA sequence AUG, coding for methionine, which initiates translation of mRNA.

Inositol lipid: A membrane-anchored phospholipid that transduces hormonal signals by stimulating the release of any of several chemical messengers.

insect resistance management: A set of strategies to reduce the frequency of resistance in crop pests. These strategies are widely referred to as Insect Resistance Management.

Insertion mutations: Changes in the base sequence of a DNA molecule resulting from the random integration of DNA from another source. See DNA, Mutation.

Insulator: A substance, object or material that does not conduct electricity. Material that conducts electricity very poorly.

Insulin: A peptide hormone secreted by the islets of Langerhans of the pancreas that regulates the level of sugar in the blood.

Integrated Circuit (IC): An electronic circuit consisting of many interconnected devices on one piece of semiconductor, typically into 10 millimeters on a side. ICs are the major building blocks of today's computers. Semiconductor circuit, typically on a very small silicon chip, containing microfabricated transistors, diodes, resistors, capacitors, etc.

Interferon: A family of small proteins that stimulate viral resistance in cells.

Intergenic regions: DNA sequences located between genes that comprise a large percentage of the human genome with no known function.

Introgression: Backcrossing of hybrids of two plant populations to introduce new genes into a wild population.

Intron: A noncoding DNA sequence within a gene that is initially transcribed into messenger RNA but is later snipped out. See Coding, DNA, Messenger RNA, Transcription.

Invasiveness: Ability of a plant to spread beyond its introduction site and become established in new locations where it may provide a deliterious effect on organisms already existing there.

Ion: A charged particle.

Isotope: One of two or more forms of an element that have the same number of protons (atomic number) but differing numbers of neutrons (mass numbers). Radioactive isotopes are commonly used to make DNA probes and metabolic tracers.

Joining (J) segment: A small DNA segment that links genes to yield a functional gene encoding an immunogobulin.

Jupiter-Brain: A posthuman being of extremely high computational power and size. This is the archetypal concentrated intelligence. The term originated due to an idea by Keith Henson that nanomachines could be used to turn the mass of Jupiter into computers running an upgraded version of himself.

Kanamycin: An antibiotic of the aminoglycoside family that poisons translation by binding to the ribosomes.

Karyotype: All of the chromosomes in a cell or an individual organism, visible through a microsope during cell division.

Kb (kilobase): The number of base pairs (in thousands) that denote the size of a nucleic acid fragment.

Kilohertz (kHz): One thousand cycles per second.

Kilojoule: Standard unit of energy equal to 1000 joules, or 0.24 kilocalories.

Lag phase: The initial growth phase, during which cell number remains relatively constant prior to rapid growth. See growth phase.

Lane: On a gel or a blot, a lane belongs to one sample. It contains fragments of different sizes that were separated by gel electrophoresis. See band, blot.

Lawn: A uniform and uninterrupted laver of bacterial growth, in which individual colonies cannot be observed.

Lectin: Protein that binds tightly to a specific sugar. Abundant lectins derived from plant seeds are often used as affinity reagents to purify glycoproteins or to detect them on the surface of cells.

Legume: A member of the pea family that possesses root nodules containing nitrogen-fixing bacteria.

Library, DNA: A large set of clones of DNA fragments from a particular organism. DNA libraries are usually maintained in *E. coli* or in bacteriophages.

Life (lifetime): The length of time the sensor can be used before its performance changes.

Life Science Applications of Deep Red T2-MP EviTag Quantum Dots: The new T2-MP EviTags are available in the deep red wavelengths used in life science research – Macoun Red (660nm) and Jonamac Red (680nm) – with carboxyl, amine, biotin and non-functionalized surfaces. Having a biologically compatible quantum dot label like the T2-MP EviTags allows further exploration and development of in vivo assays, following cells inside living organism to get a new understanding of life. In particular, it offers the possibility to create new classes of assays for use where current testing materials could produce cytotoxicity or cell death. Potential applications include live cell and even whole animal imaging, blood cancer assays, and numerous other applications.

Ligand: In protein chemistry, a small molecule that is (or can be) bound by a larger molecule is termed a ligand. In organometallic chemistry, a moiety bonded to a central metal atom is also termed a ligand; the latter definition is more common in general chemistry.

Ligase (DNA ligase): An enzyme that catalyzes a condensation reaction that links two DNA molecules via the formation of a phosphodiester bond between the 3' hydroxyl and 5' phosphate of adjacent nucleotides.

Ligate: The process of joining two or more DNA fragments.

Ligation: Joining two fragments of DNA end to end. See sticky ends.

Light Emitting Diodes (LEDs): Light Emitting Diodes work on a completely different concept. Traditionally LEDs are created from two semiconductors. By running current in one direction across the semiconductor the LED emits light of a particular frequency (hence a particular color) depending on the physical characteristics of the semiconductor used. The semiconductor is covered with a piece of plastic that focuses the light and increases the brightness. These semiconductors are very durable, there is no filament, they donít require much power, theyíre brighter and they last a long time. By densely packing red, blue and green LEDs next to each other on a substrate one can create a display.

Lineage: A chart that traces the flow of genetic information from generation to generation.

Linkage: The frequency of coinheritance of a pair of genes and/or genetic markers, which provides a measure of their physical proximity to one another on a chromosome. Measures the physical distance between two genes. Genes that are close together are unlikely to segregate in a sexual cross. Distant genes segregate independently are are then said to be unlinked.

Linked genes/markers: Genes and/or markers that are so closely associated on the chromosome that they are coinherited in 80% or more of cases.

Linker: A short, double-stranded oligonucleotide containing a restriction endonuclease recognition site, which is ligated to the ends of a DNA fragment.

Liposome: A type of nanoparticle made of lipids, or fat molecules, surrounding a water core. Liposomes, several of which are widely used to treat infectious diseases and cancer, were the first type of nanoparticle to be used to create therapeutic agents with novel characteristics. Membrane-bound vesicles constructed in the laboratory to transport biological molecules.

Liquid Crystal Display: LCD is the predominant technology used in flat panel displays. The principle that makes the display work is this: A crystalís alignment can be altered with an electric current. If the crystal is lined up one way ñ it will allow the light waves to pass through a polarized filter, but if the electric current alters the crystalís alignment, it will guide light so that the polarized filter blocks the light. By densely packing red, blue and green light emitting crystals next to each other on a sheet (ìcalled a substrateî), one can create a full color display. The great thing about LCD is that the crystals can be packed together closely, allowing for a higher-resolution, finer-detail display. The con is that LCDs are somewhat fragile, require a lot of power and are relatively less bright. Liquid Crystal

Display — display device employing light source and electrically alterable optically active thin film.

Locus (plural = loci): A specific location or site on a chromosome.

Logarithmic phase (log or exponential growth phase): The steepest slope of the growth curve—the phase of vigorous growth during which cell number doubles every 20-30 minutes.

Lysogen: A bacterial cell whose chromosome contains integrated viral DNA.

Lysogenic: A type or phase of the virus life cycle during which the virus integrates into the host chromosome of the infected cell, often remaining essentially dormant for some period of time. See Lysogen.

Lytic: A phase of the virus life cycle during which the vi- rus replicates within the host cell, releasing a new generation of viruses when the infected cell lyses.

Machine-phase chemistry: The chemistry of systems in which all potentially reactive moieties follow controlled trajectories (e.g., guided by molecular machines working in vacuum).

Macroscale: Larger than nanoscale; often implies a design that humans can directly interact with; too large to be built by a single assembler (one cubic micron of diamond contains 176 billion atoms).

Magnetization (M): The total magnetic moment per unit volume of material. Also, a measure of the contribution to the magnetic flux by some material within an H field. The magnitude of M is proportional to the applied field as: M = ÷m × H, with ÷m the magnetic susceptibility.

Magnetostrictive material: A material that changes dimension in the presence of a magnetic field or generates a magnetic field when mechanically deformed.

Manufacturing: The ability to make products, in this case ranging from clothing, to electronics, to medical devices, to books, to building materials, and much more.

Meat Machine: AKA Cabinet Beast. A box containing assemblers and raw material, within which is formed meat [or whatever else it was programmed to make].

Mechanical: Pertaining to the positions and motions of atoms, as defined by the positions of their nuclei; see electronic. A purely mechanical device can be described in terms of atomic positions and motions without reference to electronic properties, save through their effect on the potential energy function.

Mechanochemistry: Chemistry accomplished by mechanical systems directly controlling the reactant molecules; the formation or breaking of chemical bonds under direct mechanical control. In this volume, the chemistry of processes in which mechanical systems operating with atomic-scale precision either guide, drive, or are driven by chemical transformations. In general usage, the chemistry of processes in which energy is converted from mechanical to chemical form, or vice versa.

Mechanosynthesis: Chemical synthesis controlled by mechanical systems operating with atomic-scale precision, enabling direct positional selection of reaction sites; synthetic applications of mechanochemistry. Suitable mechanical systems include AFM mechanisms, molecular manipulators, and molecular mill systems. Processes that fall outside the intended scope of this definition include reactions guided by the incorporation of reactive moieties into a shared covalent framework (i.e., conventional intramolecular reactions), or by the binding of reagents to enzymes or enzymelike catalysts. Molecular tools with chemically specific tip structures can be used, sequentially, to modify a work piece and build a wide range of molecular structures.

Mechatronics: The study of the melding of AI and electromechanical machines to make machines that are greater than the sum of their parts. The synergistic combination of precision mechanical engineering with electronic control.

Megabase cloning: The cloning of very large DNA fragments.

Megabyte (MB): 220 (= 1,048,576, or about one million) bytes of information.

Megahertz (Mhz): One million cycles per second

Meiosis: The reduction division process by which haploid gametes and spores are formed, consisting of a single duplication of the genetic material followed by two mitotic divisions.

Meme: An idea that replicates through a society as it is propagated through person-to-person interaction, both direct and indirect. Memetics is a field of study that focuses on memes' role in the evolution of a culture. Self-reproducing idea or other information pattern which is propagated in ways similar to that of a gene. [Richard Dawkins, 1976]

Messenger RNA (mRNA): The class of RNA molecules that copies the genetic information from DNA, in the nucleus, and carries it to ribosomes, in the cytoplasm, where it is translated into protein.

Metabolism: The biochemical processes that sustain a living cell or organism.

Metallothionein: A protective protein that binds heavy metals, such as cadmium and lead.

Micellar nanocontainers: Block copolymer micelles are water-soluble biocompatible nanocontainers with great potential for delivering hydrophobic drugs. An understanding of their cellular distribution is essential to achieving selective delivery of drugs at the subcellular level. R. Savic, L. Luo, A. Eisenberg, D. Maysinger, Micellar nanocontainers distribute to defined cytoplasmic organelles Science 300 (5619): 615- 618, Apr. 25, 2003

Micelles: Micelles are small, spherical structures composed of molecules that attract one another to reduce surface tension. The head of the molecule is hydrophilic, meaning it likes water, while the interior portion is hydrophobic, meaning it avoids water.

Microbial mats (biofilms): Layered groups or communities of microbial populations.

Microelectronics: Narrower terms: MEMS, nanoelectronics, optoelectronics, SED Single Electron Devices. Related terms: molecular electronics, semiconductors

Microencapsulation: Individually encapsulated small particles.

Microinjection: A means to introduce a solution of DNA, protein, or other soluble material into a cell using a fine microcapillary pipet.

Microspectrophotometry: Analytical technique for studying substances present at enzyme concentrations in single cells, in situ, by measuring light absorption. Light from a tungsten strip lamp or xenon arc dispersed by a grating monochromator illuminates the optical system of a microscope. The absorbance of light is measured (in nanometers) by comparing the difference between the image of the sample and a reference image. MeSH 1990

Microspheres: Microspheres are generally defined as small spheres made of any material and sized from about 0.5 μm to 100 μm. Similar, but smaller spheres sized 10 to 500 nm are called *nanospheres.* Ideally, microspheres are completely spherical and homogeneous in size, although less perfect particles are often termed microspheres as well. Depending on the preparation method and material used, microspheres show a typical size distribution which often deviates from the mono-sized ideal.

Microstructure: In material engineering the structural features of a material such as grain boundaries, grain size and structure, subject to observation under a microscope, selective etching etc. In MEMS microstructure unfortunately is also used to designate a micromachined feature. The last decade has seen rapid developments in the fabrication, characterization and conceptual understanding of synthetic microstructures in many different material systems including silicon, III-V and II-VI semiconductors, metals, ceramics and organics. The objective of this journal [Superlattices and Microstructures] is to provide a common interdisciplinary platform for the publication of the latest research results on all such "nanostructures" with dimensions in the range of 1-100 nm; the unifying theme here being the dimensions of these artificial structures rather than the material system in which they are fabricated.

Microsystem: A microscale machine that can sense information from the environment and act accordingly. Outside the U.S., it can also refer to microelectromechanical systems (MEMS). [smalltimes glossary, 2002]

MicroTas, micro Total Analysis Systems, uTAS: Although initial research dates back to the early 1970's, the field of micro- TAS formally started in 1990, when Manz et al described the possibility of creating microsystems that would take care of many or all the traditional analytical steps involved in a biochemical analysis (sample introduction, handling, extraction, purification, concentration, filtration, analysis, detection) .... Micro-TAS offer many advantages over traditional analysis systems. Low power consumption and small reaction volumes, faster analysis, ultrasensitive detection, and minimal human intervention are key parameters in the development of micro- TAS. Most biochemical reactions take place in liquid environments. Hence, the development of MicroTAS is intrinsically linked to the design of liquid handling micro- devices. [Biomedical Applications Group (GAB) Centro Nacional de Microelectronica (CNM- IMB) Bellaterra, Spain, 2000]

Microtubule: A hollow cylindrical tube used for the transport of materials inside cells.

Miller indices: A set of 3 integers (4 for hexagonal) that designate crystallographic planes, as determined from reciprocals of fractional axial intercepts.

Millimeter: One thousandth of a meter, or about 1/26 of an inch.

Mitosis: The replication of a cell to form two daughter cells with identical sets of chromosomes.

Molecular - Nanotechnology: The technology of precisely-constructed molecular-scale machines; from nanometer: a billionth of a meter. For a good introduction, see What is Nanotechnology? or Ralph C. Merkle's nanotechnology page.

Molecular assembler: A general-purpose device for molecular manufacturing, able to guide chemical reactions by positioning individual molecules to atomic accuracy (e.g. mechanosynthesis) and to construct a wide range of useful and stable molecular structures according to precise specifications. Also known as an assembler, a molecular assembler is a molecular machine that can build a molecular structure from its component building blocks. [ZY]

Molecular Beam Epitaxy: [MBE] Process used to make compound (multi-layer) semiconductors. Consists of depositing alternating layers of materials, layer by layer, one type after another (such as the semiconductors gallium arsenide and aluminum gallium arsenide).

Molecular biology: The study of the biochemical and molecular interactions within living cells.

Molecular cloning: The biological amplification of a specific DNA sequence through mitotic division of a host cell into which it has been transformed or transfected.

Molecular genetics: The study of the flow and regulation of genetic information between DNA, RNA, and protein molecules.

Molecular markers: Genetic traits which are detectable on gels or blots and can be used to construct genetic maps. Molecular markers usually have no known function. RFLPs are molecular markers.

Molecular mechanics models: Many of the properties of molecular systems are determined by the molecular potential energy function. Molecular mechanics models approximate this function as a sum of 2-atom, 3-atom, and 4-atom terms, each determined by the geometries and bonds of the component atoms. The 2-atom and 3-atom terms describing bonded interactions roughly correspond to linear springs.

Molecular medicine: A variety of pharmaceutical techniques and gene therapies that address specific molecular diseases or molecular defects in biological systems. A variety of pharmaceutical techniques and therapies in use today. Studying molecules as they relate to health and disease, and manipulating those molecules to improve the diagnosis, prevention, and treatment of disease.

Molecular mill: A mechanochemical processing system characterized by limited motions and repetitive operations without programmable flexibility.

Molecular motors: Protein based machines that are involved in or cause movement such as the rotary devices (flagellar motor and the F1 ATPase) or the devices whose movement is directed along cytoskeletal filaments (myosin, kinesin and dynein motor families). [MeSH, 1999]

Molecular Nanoscience: An emerging interdisciplinary field that combines the study of molecular/ biomolecular systems with the science and technology of nanoscale structures and systems. The potential applications for this research are very broad and include such possibilities as 1) the use of biomolecules and cellular systems to self- assemble nanoelectronic circuitry and other nanoscale structures and 2) the use of lamellar host frameworks containing nano pores that can be tailored to include guest molecules for separation of chemicals for pharmaceutical and other applications. Molecular Nanoscience Alliance for Interdisciplinary Studies and Activities, Univ. of Minnesota, US,

Molecular Wire: A molecular wire—the simplest electronic component—is a quasi-one-dimensional molecule that can transport charge carriers (electrons or holes) between its ends. [Michael D Ward]

Molecularly imprinted polymers MIPs: A new class of materials that have artificially created receptor structures. Since their discovery in 1972, MIPs have attracted considerable interest from scientists and 3engineers involved with the development of chromatographic absorbents, membranes, sensors and enzyme and receptor mimics.

Monoclonal antibodies: Immunoglobulin molecules of single-epitope specificity that are secreted by a clone of B cells.

Monoculture: The agricultural practice of cultivating crops consisting of genetically similar organisms.

Monogenic: Controlled by or associated with a single gene.

Multi-locus probe: A probe that hybridizes to a number of different sites in the genome of an organism.

Mutagen: Any agent or process that can cause mutations. See Mutation.

Mutation: An alteration in DNA structure or sequence of a gene. A spontaneous or induced genetic change in the DNA of an organism. Most mutations are undesirable, but a few are useful such as dwarfing genes in cereal crops that allow the plants to stand better.

Mycorrhizae: Fungi that form symbiotic relationships with roots of more developed plants.

Nanobiology: Many fundamental biological functions are carried out by molecular machineries that have the sizes of 1-100 nm. You find many examples in molecular biology and cell biology: single enzymes, transcription complex, ribosome, transport complex, nuclear pore, and so on. To understand the functions of these machineries, one has to describe their movements, changes in their shapes, and their localization. This means the mechanistic study is equivalent to dynamic morphology at this level of size, making a new field that merges mechanistic biology and morphology. The emergence of nanobiology depended on the invention of nano- technology: scanning probe microscopy, modern optical techniques, and micro- manipulating techniques. This concept of nanobiology was first proposed in a group study named "Biological Nano- Mechanisms", which was supported by Japanese Agency of Science and Technology (1992-1997). [National Institute of Genetics, Japan]

Nanobiotechnology: An emerging area of scientific and technological opportunity. Nanobiotechnology applies the tools and processes of nano/ microfabrication to build devices for studying biosystems. Researchers also learn from biology how to create better micro- nanoscale devices. The Nanobiotechnology Center (NBTC), a National Science Foundation, Science and Technology Center is characterized by its highly interdisciplinary nature and features a close collaboration between life scientists, physical scientists, and engineers. [NBTC, NanoBiotechnology Center, Cornell Univ. US, 2002] Use of nanotechnology[ies] in the life sciences. These may include, but are not limited to, therapeutics, medical devices/implants, biosensors, and tools for the development of drugs. Applying the tools and processes of MNT to build devices for studying biosystems, in order to learn from biology how to create better nanoscale devices. Should hasten the creation of useful micro devices that mimic living biological systems.

Nanocentrifuge: In medical nanorobotics, a proposed nanodevice that can spin materials at very high speed, imparting rotational accelerations of up to one trillion gravities (g's), thus permitting rapid sortation.

Nanoscience: Nanoscience is the science of nanoscale phenomena. This is a very complicated area, since Quantum Mechanics is often needed to describe the physical properties of atoms and molecules properly. What makes it even more complicated is that nanostructured materials can still be quite a challenge to describe Quantum Mechanically, since the equations are feindishly difficult—this is where supercomputers come in to make the job approachable. The scientific discipline seeking to increase our knowledge and understanding of nanoscale phenomena, i.e. science on the scale of 0.1 nm to 100 nm.

Nanosensor: A chemical or physical sensor constructed using nanoscale components, usually microscopic or submicroscopic in size.

Nanosources: sources that emit light from nanometre-scale volumes.

Nanotech: Slang for nanotechnology.

National Institutions of Health (NIH): A nonregulatory agency which has oversight of research activities that the agency funds.

National Science Foundation (NSF): A nonregulatory agency which has oversight of biotechnology research activities that the agency funds.

Natural selection: The differential survival and reproduction of organisms with genetic characteristics that enable them to better utilize environmental resources.

Neophile: One who welcomes the future and who enjoys change and evolution.

Nick translation: A procedure for making a DNA probe in which a DNA fragment is treated with DNase to produce single-stranded nicks, followed by incorporation of radioactive nucleotides from the nicked sites by DNA polymerase I.

Nicked circle (relaxed circle): During extraction of plasmid DNA from the bacterial cell, one strand of the DNA becomes nicked. This relaxes the torsional strain needed to maintain supercoiling, producing the familiar form of plasmid.

Nitrocellulose: A membrane used to immobilize DNA, RNA, or protein, which can then be probed with a labeled sequence or antibody.

Nitrogen fixation: The conversion of atmospheric nitrogen to biologically usable nitrates.

Nitrogenous bases: The purines (adenine and guanine) and pyrimidines (thymine, cytosine, and uracil) that comprise DNA and RNA molecules.

Nodule: The enlargement or swelling on roots of nitrogen-fixing plants. The nodules contain symbiotic nitrogen-fixing bacteria. See Nitrogen fixation.

Nontarget organism: An organism which is affected by an interaction for which it was not the intended recipient.

Northern hybridization: (Northern blotting). A procedure in which RNA fragments are transferred from an agarose gel to a nitrocellulose filter, where the RNA is then hybridized to a radioactive probe.

Novel foods: Novel foods are products that have never been used as a food; foods which result from a process that has not previously been used for food; or, foods that have been modified by genetic manipulation. This last category of foods have been described as genetically modified foods (often referred to as GM foods, genetically engineered foods or biotechnology-derived foods). *Definition from Health Canada Web site*

Nuclease: Any enzyme that cuts nucleic acids. See restriction enzyme. A class of enzymes that degrades DNA and/or RNA molecules by cleaving the phosphodiester bonds that link adjacent nucleotides. In deoxyribonuclease (DNase), the substrate is DNA. In endonuclease, it cleaves at internal sites in the substrate molecule. Exonuclease progressively cleaves from the end of the substrate molecule. In ribonuclease (RNase), the substrate is RNA. In the S1 nuclease, the substrate is single-stranded DNA or RNA.

Nucleic acids: The two nucleic acids, deoxyribonucleic acid (DNA) and ribonucleic acid (RNA), are made up of long chains of molecules called nucleotides. See DNA, RNA, Nucleotides.

Nuclein: The term used by Friedrich Miescher to describe the nuclear material he discovered in 1869, which today is known as DNA.

Nucleoside analog: A synthetic molecule that resembles a naturally occuring nucleoside, but that lacks a bond site needed to link it to an adjacent nucleotide.

Nucleotide: A building block of DNA and RNA, consisting of a nitrogenous base, a five-carbon sugar, and a phosphate group. Together, the nucleotides form codons, which when strung together form genes, which in turn link to form chromosomes.

Nucleus: The membrane-bound region of a eukaryotic cell that contains the chromosomes. A cellular organelle in plants and animals that contains the chromosomes which in turn are composed of DNA plus protein.

Occupational Safety and Health Administration (OSHA): One of the U.S. agencies responsible for regulation of biotechnology. The major law under which the agency has regulatory powers is the Occupational Safety and Health Act.

Oligonucleotide: A DNA polymer composed of only a few nucleotides.

Oncogene: A gene that contributes to cancer formation when mutated or inappropriately expressed.

Oncogenesis: The progression of cytological, genetic, and cellular changes that culminate in a malignant tumor.

Open pollination: Pollination by wind, insects, or other natural mechanisms.

Open reading frame: A long DNA sequence that is uninterrupted by a stop codon and encodes part or all of a protein. (

Operator: A prokaryotic regulatory element that interacts with a repressor to control the transcription of adjacent structural genes.

Orbital: In the approximation that each electron in a molecule has a distinct, independent wave function, the spatial distribution of an electron wave function corresponds to a molecular orbital. These, in turn, can be approximated as sums of contributions from the orbitals characteristic of the isolated atoms. An electron added to a molecule—or, similarly, one excited to a higher-energy state within a molecule—would occupy a state with a different wave function from the rest; an unoccupied state of this kind corresponds to an unoccupied molecular orbital. Orbital-symmetry effects on reaction rates arise when a reaction requires overlap between two lobes of the orbitals on each of two reagents: if the algebraic signs of the wave functions in the facing lobes do not match, bond formation between those orbitals is prohibited.

Organelle: A cell structure that carries out a specialized function in the life of a cell.

Origin of replication: The nucleotide sequence at which DNA synthesis is initiated.

Overlapping reading frames: Start codons in different reading frames generate different polypeptides from the same DNA sequence.

Ovum: A female gamete.

Paleontology: The study of the fossil record of past geological periods and of the phylogenetic relationships between ancient and contemporary plant and animal species.

Palindromic sequence: A DNA locus whose 5'-to-3' sequence is identical on each DNA strand. The sequence is the same when one strand is read left to right and the other strand is read right to left. Recognition sites of many restriction enzymes are palindromic. See DNA.

pAMP: Ampicillin-resistant plasmid developed for this laboratory course.

Parasitism: The closee association of two or more dissimilar organisms where the association is harmful to at least one. See Commensalism, Parasitism, Symbiosis.

Particle bombardment: Using metal particles to blast DNA into cells for the purpose of genetic engineering. A gene gun is used to propel the metal particles.

Pathogen: Organism which can cause disease in another organism.

pBR322: A derivation of ColE1, one of the first plasmid vectors widely used.

PCR: Polymerase chain reaction. An amazingly sensitive technique that allows the specific amplification of extremely small amounts of particular DNA fragments using DNA polymerase and specific primers.

Pedigree: A diagram mapping the genetic history of a particular family.

Percent Accepted Mutation: A unit introduced by Dayhoff et al. to quantify the amount of evolutionary change in a protein sequence. 1.0 PAM unit, is the amount of evolution which will change, on average, 1% of amino acids in a protein sequence. A PAM(x) substitution matrix is a look-up table in which scores for each amino acid substitution have been calculated based on the frequency of that substitution in closely related proteins that have experienced a certain amount (x) of evolutionary divergence.

Persistence: Ability of an organism to remain in a particular setting for a period of time after it is introduced.

Pesticide: A substance that kills harmful organisms (for example, an insecticide or fungicide).

Phenotype: The observable characteristics of an organism, the expression of gene alleles (genotype) as an observable physical or biochemical trait. See Genotype.

Pheromone: A hormone-like substance that is secreted into the environment.

Phosphatase: An enzyme that hydrolyzes esters of phosphoric acid, removing a phosphate group.

Phosphodiester bond: A bond in which a phosphate group joins adjacent carbons through ester linkages. A condensation reaction between adjacent nucleotides results in a phosphodiester bond between 3' and 5' carbons in DNA and RNA.

Phospholipid: A class of lipid molecules in which a phosphate group is linked to glycerol and two fatty acyl groups. A chief component of biological membranes.

Phosphorylation: The addition of a phosphate group to a compound.

Physical map: A map showing physical locations on a DNA

molecule, such as restriction sites, and sequence-tagged sites.

Piezoelectric material: A ferroelectric material in which an electrical potential difference is created due to mechanical deformation, or conversely, in which the application of a voltage causes dimensional changes in the material.

Piezoelectric: A piezoelectric material expands or contracts when a voltage is applied across it. Piezoelectric tubes are often used to move the tip in an AFM or STM. If a piezoelectric crystal is compressed, a voltage appears across it. This effect is sometimes used to measure displacements. A commonly used piezoelectric is PZT (Pb,Zr)TiO3.

Plaque: A clear spot on a lawn of bacteria or cultured cells where cells have been Iysed by viral infection.

Plasmid (p): A circular DNA molecule, capable of autonomous replication, which typically carries one or more genes encoding antibiotic resistance proteins. Plasmids can transfer genes between bacteria and are important tools of transformation for genetic engineers.

Pleiotrophy: The effect of a particular gene on several different traits.

Point mutation: A change in a single base pair of a DNA sequence in a gene.

Poly(A) polymerase: Catalyzes the addition of adenine residues to the 3' end of pre-mRNAs to form the poly(A) tail.

Polyacrylamide gel electrophoresis: Electrophoresis through a matrix composed of a synthetic polymer, used to separate proteins, small DNA, or RNA molecules of up to 1000 nucleotides. Used in DNA sequencing.

Polyclonal antibodies: A mixture of immunoglobulin molecules secreted against a specific antigen, each recognizing a different epitope.

Polygenic: Controlled by or associated with more than one gene.

Polylinker: A short DNA sequence containing several restriction enzyme recognition sites that is contained in cloning vectors.

Polymer: A molecule composed of repeated subunits.

Polymerase (DNA): Synthesizes a double-stranded DNA molecule using a primer and DNA as a template.

Polymerase chain reaction (PCR): A procedure that enzymatically amplifies a DNA polymerase.

Polymerase: Any enzyme that adds subunits to chains of macromolecules. Example is DNA polymerase.

Polymorphisms: Variant forms of a particular gene that occur simultaneously in a population.

Polynucleotide: A DNA polymer composed of multiple nucleotides.

Polypeptide (protein): A polymer composed of multiple amino acid units linked by peptide bonds.

Polyploid: A multiple of the haploid chromosome number that results from chromosome replication without nuclear division.

Polysaccharide: A polymer composed of multiple units of monosaccharide (simple sugar).

Polyvalent vaccine: A recombinant organism into which has been cloned antigenic determinants from a number of different disease-causing organisms.

Population: A local group of organisms belonging to the same species and capable of interbreeding.

Primary cell: A cell or cell line taken directly from a living organism, which is not immortalized.

Primer: A short DNA or RNA fragment annealed to single-stranded DNA, from which DNA polymerase extends a new DNA strand to produce a duplex molecule.

Prion: See Proteinaceous infectious particle.

Probe: A single-stranded DNA that has been radioactively labeled and is used to identify complementary sequences in genes or DNA fragments of interest. A fragment of DNA, usually labeled with radioactive 32P, that detects homologous sequences on southern blots.

Prokaryote: A bacterial cell lacking a true nucleus; its DNA is usually in one long strand. See Eukaryote.

Promoter: A region of DNA extending 150-300 bp upstream from the transcription start site that contains binding sites for RNA polymerase and a number of proteins that regulate the rate of transcription of the adjacent gene.

Pronucleus: Either of the two haploid gamete nuclei just prior to their fusion in the fertilized ovum.

Protease: An enzyme that cleaves peptide bonds that link amino acids in protein molecules.

Protein: Molecules composed of amino acids. Proteins constitute the enzymes and many of the structural components of cells. A polymer of amino acids linked via peptide bonds and which may be composed of two or more polypeptide chains. Proteins do the work in cells. They can be part of structures (such as cell walls, organelles, etc), regulate reactions that take place in the cell or they can serve as enzymes, which speed-up reactions. Everything you see in an organism is either made of proteins or the result of a protein action.

Protein kinase: An enzyme that adds phosphate groups to a protein molecule at serine, threonine, or tyrosine residues.

Proteinaceous infectious particle (prion): A proposed pathogen composed only of protein with no detectable nucleic acid and which is responsible for Creutzfeldt-Jakob disease and kuru in humans and scrapie in sheep.

Quantum Mechanics: A largely computational physical theory explaining the behavior of quantum phenomena, which incorporates the theory of special relativity. Despite dilignet attempts, general relativity has not been sucessfully incorporated into quantum mechanics. A physical model of chemical and optical phenomena, as well as the behaviour of matter in general on a small scale. Quantum mechanics describes a system of particles in terms of a wave function defined over the configuration space of the system. Although the concept of particles having distinct locations is implicit in the potential energy function that determines the wave function (e.g., of a ground-state system), the observable dynamics of the system cannot be described in terms of the motion of such particles from point to point. In describing the energies, distributions, and behaviors of electrons in nanometer-scale structures, quantum mechanical methods are necessary. Electron wave functions help determine the potential energy surface of a molecular system, which in turn is the basis for classical descriptions of molecular motion. Nanomechanical systems can almost always be described in terms of classical mechanics, with occasional quantum mechanical corrections applied within the framework of a classical model.

Quantum Well: A P-N-P junction in which the "N" layer is ~10 nm (where traditional physics leaves off and quantum effects take over) and an "electron trap" is created. "If one makes a heterostructure with sufficiently thin layers, quantum interference effects begin to appear prominently in the motion of the electrons. The simplest structure in which these may be observed is a quantum well, which simply consists of a thin layer of a narrower-gap semiconductor between thicker layers of a wider-gap material."

rDNA: Def. 1. Ribosomal DNA; DNA which codes for ribosomes

Reading frame: A series of triplet codons beginning from a specific nucleotide. Depending on where one begins, each DNA strand contains three different reading frames.

Recessive gene: Characterized as having a phenotype expressed only when both copies of the gene are mutated or missing.

Recessive(-acting) oncogene, (anti-oncogene): A single copy of this gene is sufficient to suppress cell proliferation; the loss of both copies of the gene contributes to cancer formation.

Recessive: Refers to the member of a pair of alleles that fails to be expressed in the phenotype of the organism when the dominant member is present. Also refers to the phenotype of an individual that has only the recessive allele.

Recognition sequence (site): A nucleotide sequence—composed typically of 4, 6, or 8 nucleotides—that is recognized by a restriction endonuclease. Type II enzyrnes cut (and their corresponding modification enzymes methylate) within or very near the recognition sequence.

Recombinant DNA: The process of cutting and recombining DNA fragments from different sources as a means to isolate genes or to alter their structure and function.

Recombination frequency: The frequency at which crossing over occurs between two chromosomal loci—the probability that two loci will become unlinked during meiosis.

Refuge: A refuge is any host plant (non-Bt crop, potatoes, oats, sorghum, and some weeds) not producing Bt proteins or not being treated with conventional Bt formulations. The purpose of the refuge is to supply a source of susceptible insects that may mate with resistant insects of the same species emerging from nearby Bt crop fields. If the refuge is large enough, susceptible insect numbers should outnumber resistant individuals, and in theory, a greater number of the next generation offspring will be susceptible to the Bt crop.

Regulatory gene: A gene whose protein controls the activity of other genes or metabolic pathways.

Relaxed plasmid: A plasmid that replicates independently of the main bacterial chromosome and is present in 10-500 copies per cell.

Renature: The reannealing (hydrogen bonding) of single-stranded DNA and/or RNA to form a duplex molecule.

Replicon: A chromosomal region containing the DNA sequences necessary to initiate DNA replication processes.

Repressor: A DNA-binding protein in prokaryotes that blocks gene transcription by binding to the operator.

Resistance: The evolved ability of a species to survive during a stressful event or set of circumstances.

Restriction enzyme: A class of enzymes that cut DNA at specific sequences called restriction sites. Restriction enzymes were key to making genetic engineering possible.

Restriction-fragment-length polymorphism (RFLP): Differences in nucleotide sequence between alleles at a chromosomal locus result in restriction fragments of varying lengths detected by Southern analysis.

Retrovirus: A member of a class of RNA viruses that utilizes the enzyme reverse transcriptase to reverse copy its genome into a DNA intermediate, which integrates into the hostcell chromosome. Many naturally occurring cancers of vertebrate animals are caused by retroviruses.

Reverse genetics: Using linkage analysis and polymorphic markers to isolate a disease gene in the absence of a known metabolic defect, then using the DNA sequence of the cloned gene to predict the amino acid sequence of its encoded protein.

Reverse transcriptase (RNA-dependent DNA polymerase): An enzyme isolated from retrovirus-infected cells that synthesizes a complementary (c)DNA strand from an RNA template.

RFLP: Restriction fragment length polymorphism. RFLPs are generated by digesting DNA with restriction enzymes. The DNA is separated by size by gel electrophoresis. Slight differences in homologous fragments may exist between individuals. These length polymorphisms are RFLPs. See blot.

Rhizobia: Bacteria in a symbiotic relationship with leguminous plants that results in nitrogen fixation. See Nitrogen fixation.

Rhizosphere: The soils region on and around plant roots.

Ribosomal RNA (rRNA): The RNA component of the ribosome.

Ribosome: Cellular organelle that is the site of protein synthesis during translation. See Organelle, Translation.

Ribosome-binding site: The region of an mRNA molecule that binds the ribosome to initiate translation.

Risk assessment: Defined as a formalized basis for the objective evaluation of risk in a manner in which assumptions and uncertainties are clearly considered and presented.

RNA (ribonucleic acid): An organic acid composed of repeating nucleotide units of adenine, guanine, cytosine, and uracil, whose ribose components are linked by phosphodiester bonds.

RNA polymerase: Transcribes RNA from a DNA template.

RNA: Ribonucleic acid. The three basic types are messenger RNA (mRNA), ribosomal RNA (rRNA), and transfer RNA (tRNA).

Sequence: Noun: the particular order of nucleotides in a DNA or RNA fragment. Verb: to determine the particular order of nucleotides in a strand of DNA.

Sequence-tagged site (STS): A unique (single-copy) DNA sequence used as a mapping landmark on a chromosome.

Sexual reproduction: The process where two cells (gametes) fuse to form one hybrid, fertilized cell. See Asexual reproduction, Gamete, Hybrid.

Signal transduction: The biochemical events that conduct the signal of a hormone or growth factor from the cell exterior, through the cell membrane, and into the cytoplasm. This involves a number of molecules, including receptors, pro- teins, and messengers.

Site-directed mutagenesis: The process of introducing specific base-pair mutations into a gene.

Small nuclear RNA (snRNA): Short RNA transcripts of 100-300 bp that associate with proteins to form small nuclear ribonucleoprotein particles (snRNPs), which participate in RNA processing.

Somatic cell gene therapy: The repair or replacement of a defective gene within somatic tissue.

Somatic cell: Any nongerm cell that composes the body of an organism and which possesses a set of multiploid chromosomes (diploid in most organisms).

Southern hybridization (Southern blotting): A procedure in which DNA restriction fragments are transferred from an agarose gel to a nitrocellulose filter, where the denatured DNA is then hybridized to a radioactive probe (blotting).

Species: A classification of related organisms that can freely interbreed.

Spore: A form taken by certain microbes that enables them to exist in a dormant stage. It is an asexual reproductive cell. See Asexual reproduction, Dormant.

of topography and electronic properties. The STM has been used to manipulate atoms and molecules on surfaces. A type of scanning probe microscopy that maps the local electrical properties of an interface.

Schwann cell: Glial cell responsible for forming myelin sheaths in the peripheral nervous system.

Science: The process of developing a systematized knowledge of the world through the variation and testing of hypotheses. The pursuit of knowledge and understanding, from the Latin term *scientia*, which means 'knowledge'.

Selectable marker genes: Genetic engineers must be able to select cells that have received the transgene from those that have not. Selecting out transgenic cells is done by co-transforming the cells with the transgene plus an additional gene called a selectable marker gene. Selectable marker genes are genes that encode easily detectable traits making transgenic cells discernible from non-transgenic cells. The two most commonly used selectable marker genes encode the traits of herbicide and antibiotic resistance.

Selectable marker: A gene whose expression allows one to identify cells that have been transforrned or transfected with a vector containing the marker gene.

Self-pollination: Pollen of one plant is transferred to the female part of the same plant or another plant with the same genetic makeup.

Semiconservative replication: During DNA duplication, each strand of a parent DNA molecule is a template for the synthesis of its new complementary strand. Thus, one half of a preexisting DNA molecule is conserved during each round of replication.

Sequence hypothesis: Francis Crick's seminal concept that genetic information exists as a linear DNA code; DNA and protein sequence are colinear.

Roundup Ready: A trademark for plants that are genetically engineered to be resistant to the herbicide Roundup (technical name: glyphosate).

Salmonella: A genus of rod-shaped, gram-negative bacteria that are a common cause of food poisoning.

Satellite RNA (viroids): A small, self-splicing RNA molecule that accompanies several plant viruses, including tobacco ringspot virus.

Scanning Near Field Optical Microscopy: A method for observing local optical properties of a surface that can be smaller than the wavelength of the light used. A type of scanning probe microscopy that maps the near-field optical properties of an interface.

Scanning Probe Microscopy (SPM): A method for imaging nanoscale features of surfaces by scanning a sensor (probe) over a surface. Near-field effects such as tunneling, van der Waals forces, local fields and more are serially detected at localized points on the surface and used to create an SPM image.

Scanning Thermal Microscopy (SThM): A type of scanning probe microscopy that maps the local temperature and thermal conductivity of an interface. A method for observing local temperatures and temperature gradients on a surface. [NTN]

Scanning Tunneling Microscope (STM): An instrument able to image conducting surfaces to atomic accuracy; has been used to pin: molecules to a surface. A device in which a sharp conductive tip is moved across a conductive surface close enough to permit a substantial tunneling current (typically a nanometer or less). In a common mode of operation, the voltage is kept constant and the current is monitored and kept constant by controlling the height of the tip above the surface; the result, under favorable conditions, is an atomic-resolution map of the surface reflecting a combination

Stationary phase: The plateau of the growth curve after log growth, during which cell number remains constant. New cells are produced at the same rate as older cells die.

Sticky end: A protruding, single-stranded nucleotide se- quence produced when a restriction endonuclease cleaves off center in its recognition sequence. After cutting with restriction enzymes, the ends of DNA fragments are "sticky". They can easily be joined with other fragments that were cut with the same enzyme and so have complementary sticky ends.

Stringency: Reaction conditions—notably temperature, salt, and pH—that dictate the annealing of single-stranded DNA/DNA, DNA/RNA, and RNA/RNA hybrids. At high stringency, duplexes form only between strands with perfect one-to-one complementarity; lower stringency allows annealing between strands with some degree of mismatch between bases.

Stringent plasmid: A plasmid that only replicates along with the main bacterial chromosome and is present as a single copy, or at most several copies, per cell.

Structure-functionalism: The scientific tradition that stresses the relationship between a physical structure and its function, for example, the related disciplines of anatomy and physiology.

Subcloning: The process of tranferring a cloned DNA fragment from one vector to another.

Substantial equivalence: The concept of substantial equivalence is used as a guide in the safety assessment of genetically modified foods by comparing the novel food to its unmodified counterpart which has a history of safe use. This approach allows regulatory agencies to include in their consideration, the substantial history of information related to foods which have long been safely consumed in the human diet to aid in the identification of potential safety and nutritional issues. *Definition from Health Canada Web site.*

Subunit vaccine: A vaccine composed of a purified antigenic determinant that is separated from the virulent organism.

Supercoiled plasmid: The predominant in vivo form of plasmid, in which the plasmid is coiled around histone-like proteins. Supporting proteins are stripped away during extraction from the bacterial cell, causing the plasmid molecule to supercoil around itself in vitro.

Supergene: A group of neighboring genes on a chromosome that tend to be inherited together and sometimes are functionally related.

Supernatant: The soluble liquid &action of a sample after centrifugation or precipitation of insoluble solids.

Susceptibility: When individual or species will become ill or possibly die in the face of the stressful event or circumstances.

Symbiosis: The close association of two or more dissimilar organisms where both receive an advantage from the association. See Commensalism, Parasitism.

Synapsis: The pairing of homologous chromosome pairs during prophase of the first meiotic division, when crossing over occurs.

Taq polymerase: A heat-stable DNA polymerase isolated from the bacterium Therrnus aquaticus, used in PCR.

TATA box: An adenine- and thymine-rich promoter sequence located 25-30 bp upstream of a gene, which is the binding site of RNA polymerase.

T-DNA (transfer DNA, tumor-DNA): The transforming region of DNA in the Ti plasmid of Agrobacterium tumefaciens.

Template: An RNA or single-stranded DNA molecule upon which a complementary nucleotide strand is synthesized.

Termination codon: Any of three mRNA sequences (UGA,

UAG, UAA) that do not code for an amino acid and thus signal the end of protein synthesis. Also known as stop codon.

Terminator region: A DNA sequence that signals the end of transcription.

Tesla [T]: Unit of magnetic induction: 1T = 1 weber/m2 (also, 1T = 104 gauss).

Thermistor: A temperature-measuring device, that contains a resistor or semiconductor whose resistance varies with temperature.

Thermocouple: A temperature-measuring device, which contains a pair of end-joined dissimilar conductors in which an electromotive force is developed by thermoelectric effects when the joined ends and the free ends of the conductors are a different temperature.

Thiol: An SH group, or a molecule containing one. Also known as a sulfhydryl or mercapto group.

Threshold: The smallest input signal that will cause a readable change in the output signal.

Thrombin : The enzyme that converts *fibrinogen* to *fibrin* to form a blood clot. Thrombin circulates in the inactive form, prothrombin.

Thylakoid: Flattened sac of membrane in a chloroplast that contains pigment and carries out the light-gathering reactions of photosynthesis. Stacks of thylakoids form the grana of chloroplasts.

Thymidine kinase (tk): An enzyme that allows a cell to utilize an alternate metabolic pathway for incorporating thymidine into DNA. Used as a selectable marker to identify transfected eukaryotic cells.

Ti (tumor-inducing) plasmid: A giant plasmid of Agrobacterium tumefaciens that is responsible for tumor formation in infected plants. Ti plasmids are used as vectors to introduce foreign DNA into plant cells.

Toxic Substances Control Act (TSCA): See Environmental Protection Agency.

Transcapsidation: The partial of full coating of the nucleic acid of one virus with a coat protein of a differing virus. See Coat protein.

Tanscription: The copying of genetic information from DNA to messenger RNA (mRNA).

Transduction: The transfer of DNA sequences from one bacterium to another via lysogenic infection by a bacteriophage (transducing phage).

Transgenic: An organism that has incorporated a functional foreign gene through recombinant DNA technology. The novel gene exists in all of its cells and is passed through to progeny. Same as genetically modified organism (GMO).

Transition-state intermediate: In a chemical reaction, an unstable and high-energy configuration assumed by reactants on the way to making products. Enzymes are thought to bind and stabilize the transition state, thus lowering the energy of activation needed to drive the reaction to completion.

Translation: The process of converting the genetic information of an mRNA on ribosomes into a polypeptide. Transfer RNA molecules carry the appropriate amino acids to the ribosome, where they are joined by peptide bonds.

Translocation: The movement or reciprocal exchange of large-chromosomal segments, typically between two different chromosomes.

Transposition: The movement of a DNA segment within the genome of an organism.

Transposon (transposable, or movable genetic element): A relatively small DNA segment that has the ability to move from one chromosomal position to another.

tRNA (transfer RNA): The class of small RNA molecules that transfer amino acids to the ribosome during protein synthesis. See Transfer RNA.

Trypsin: A proteolytic enzyme that hydrolyzes peptide bonds on the carboxyl side of the amino acids arginine and lysine.

TSCA: The Toxic Substances Control Act. See Environmental Protection Agency.

Tumor virus: A virus capable of transforming a cell to a malignant phenotype.

U:S: Department of Agriculture: The U.S. agency responsible for regulation of biotechnology products in plants and animals. The major laws under which the agency has regulatory powers include the Federal Plant Pest Act (PPA), the Federal Seed Act, and the Plant Variety Act (PVA). In addition, the Science and Education (S&E) division has nonregulatory oversight of research activities that the agency funds.

Upload: (a) To transfer the consciousness and mental structure of a person from a biological matrix to an electronic or informational matrix (this assumes that the strong AI postulate holds). The term "Downloading" is also sometimes used, mainly to denote transferring the mind to a slower or less spacious matrix. (b) The resulting infomorph person. [The origin of the term is uncertain, but obviously based on the computer technology term 'uploading' (loading data into a mainframe computer).] [AS]

Upstream: The region extending in a 5' direction from a gene.

Vaccine: A preparation of dead or weakened pathogen, or of derived antigenic determinants, that is used to induce formation of antibodies or immunity against the pathogen.

Vaccinia: The cowpox virus used to vaccinate against smallpox and, experimentally, as a carrier of genes for antigenic determinants cloned from other disease organisms.

Variable surface glycoprotein (VSG): One of a battery of antigenic determinants expressed by a microorganism to elude immune detection.

Variation: Differences in the frequency of genes and traits among individual organisms within a population.

Vasculocyte: In medical nanorobotics, a theorized (nanorobotic) device capable of performing repairs of an injured vascular luminal surface.

Vasculography: A physical description (and mapping) of the human vascular system.

VC (Venture capitalist): A person or organisation that invests in promising companies during early development stages.

Vector: An autonomously replicating DNA molecule into which foreign DNA fragments are inserted and then propagated in a host cell. Also living carriers of genetic material (such as pollen) from plant to plant, such as insects.

Vector: In cell biology, an agent (virus or plasmid) used to transmit genetic material to a cell or organism.

Vegetal pole: The end at which most of the yolk is located in an animal egg. The end opposite the animal pole.

Viral oncogene: A viral gene that contributes to malignancies in vertebrate hosts.

Viroid: A plant pathogen that consists of a naked RNA molecule of approximately 250-350 nucleotides, whose extensive base pairing results in a nearly correct double helix.

Virtual Nanomedicine: Using VR to perform surgery and other functions inside the body.

Virtual reality system: A combination of computer and interface devices (goggles, gloves, etc.) that presents a user with the illusion of being in a three dimensional world of computer-generated objects.

Virulence: The degree of ability of an organism to cause disease.

Virus: A parasite (consisting primarily of genetic material) that invades cells and takes over their molecular machinery in order to copy itself. A small replicator consisting of little but a package of DNA or RNA which, when injected into a host cell, can direct the cell's molecular machinery to make more viruses. Particle consisting of nucleic acid (RNA or DNA) enclosed in a protein coat and capable of replicating within a host cell and spreading from cell to cell. Often the cause of disease.

Viscoelasticity: A type of deformation exhibiting the mechanical characteristics of viscous flow and elastic deformation.

Viscosity: The ratio of the magnitude of an applied shear stress to the velocity gradient that it produces; in other words: a measure of a noncrystalline material's resistance to permanent deformation.

1

Agricultural Biotechnology: An Overview

Science and technology helped revolutionize agriculture in the 20th century in many parts of the world. This issue of *Economic Perspectives* highlights how advances in biotechnology can be adapted to benefit the world in the 21st century, particularly developing countries.

Increasing yield potential and desirable traits in plant and animal food products has long been a goal of agricultural science. That is still the goal of agricultural biotechnology, which can be an important tool in reducing hunger and feeding the planet's expanding and longer-living population, while reducing the adverse environmental effects of farming practices.

In a supportive policy and regulatory environment, biotechnology has enormous potential to create crops that resist extreme weather, diseases and pests; require fewer chemicals; and are more nutritious for the humans and livestock that consume them. But there is also controversy surrounding this new technology. The journal addresses the controversies head on and provides sound scientific reasoning for the use of this technology.

In June 2003, agriculture, health and environment

ministers from over 110 countries gathered in California and learned first hand how technology, including biotechnology, can increase productivity and reduce global hunger. By sharing information on how technology can increase agricultural productivity, we can help alleviate world hunger.

Contributors to this journal include Under Secretary of State Alan Larson, Under Secretary of Agriculture J.B. Penn, Deputy Food and Drug Administration Commissioner Lester Crawford, and Ambassador Tony Hall, U.S. Representative to the U.N. Agencies for Food and Agriculture, who address a broad range of topics from the basic science of biotechnology to food safety and labeling issues. Their articles are complemented by essays from an internationally respected group of researchers and academics, a State Department fact sheet on the Cartagena Biosafety Protocol and additional resource information.

Millions of farmers worldwide now plant biotech crops, agency says:

> The U.S. Agency for International Development (USAID) has issued a series of six fact sheets outlining the key issues related to products derived from modern biotechnology. The first fact sheet defines agricultural biotechnology.

For about 10,000 years, farmers have been improving wild plants and animals through the selection and breeding of desirable characteristics. This breeding has resulted in the domesticated plants and animals that are commonly used in crop and livestock agriculture. In the twentieth century, breeding became more sophisticated, as the traits that breeders select for include increased yield, disease and pest resistance, drought resistance and enhanced flavor. Traits are passed from one generation to the next through genes, which are made of DNA. All living things — including the fruits, vegetables and meat that we eat — contain genes that tell cells how to function. Recently, scientists have learned enough to begin to identify and work with the genes (DNA) that are responsible for traits.

Agricultural biotechnology is a collection of scientific techniques used to improve plants, animals and microorganisms. Based on an understanding of DNA, scientists have developed solutions to increase agricultural productivity. Starting from the ability to identify genes that may confer advantages on certain crops, and the ability to work with such characteristics very precisely, biotechnology enhances breeders' ability to make improvements in crops and livestock. Biotechnology enables improvements that are not possible with traditional crossing of related species alone.

Genetic engineering: Scientists have learned how to move genes from one organism to another. This has been called genetic modification (GM), genetic engineering (GE) or genetic improvement (GI). Regardless of the name, the process allows the transfer of useful characteristics (such as resistance to a disease) into a plant, animal or microorganism by inserting genes (DNA) from another organism. Virtually ail crops improved with transferred DNA (often called GM crops or GMOs) to date have been developed to aid farmers to increase productivity by reducing crop damage from weeds, diseases or insects.

Molecular markers: Traditional breeding involves selection of individual plants or animals based on visible or measurable traits. By examining the DNA of an organism, scientists can use molecular markers to select plants or animals that possess a desirable gene, even in the absence of a visible trait. Thus, breeding is more precise and efficient. For example, the International Institute of Tropical

Agriculture has used molecular markers to obtain cowpea resistant to bruchid (a beetle), disease-resistant white yam and cassava resistant to Cassava Mosaic Disease, among others. Another use of molecular markers is to identify undesirable genes that can be eliminated in future generations.

Molecular diagnostics: Molecular diagnostics are methods to detect genes or gene products that are very precise and specific. Molecular diagnostics are used in agriculture to more accurately diagnose crop/livestock diseases.

Vaccines: Biotechnology-derived vaccines are used in livestock and humans. They may be cheaper, better and/or safer than traditional vaccines. They are also stable at room temperature, and do not need refrigerated storage; this is an important advantage for smallholders in tropical countries. Some are new vaccines, which offer protection for the first time against some infectious illnesses. For example, in the Philippines, biotechnology has been used to develop an improved vaccine to protect cattle and water buffalo against hemorrhagic septicemia, a leading cause of death for both species.

Tissue culture: Tissue culture is the regeneration of plants in the laboratory from disease-free plant parts. This technique allows for the reproduction of disease-free planting material for crops. Examples of crops produced using tissue culture include citrus, pineapples, avocados, mangoes, bananas, coffee and papaya.

The first food product of biotechnology (an enzyme used in cheese production and a yeast used for baking) appeared on the market in 1990. Since 1995, farmers have been growing GE crops. In 2003, 7 million farmers in 18 countries — more than 85 percent of them resource-poor farmers in the developing world — were planting biotech crops. Almost one third of the global biotech crop area was grown in developing countries.

A safe and sufficient food supply, grown in an environmentally responsible fashion is essential for humanity. Like any technology, agricultural biotechnology will have economic and social impacts. Since their introduction, crops improved using biotechnology have been used safely, with benefits such as the reduction of pesticide use. Agricultural biotechnology is only one factor among many influencing the health and welfare of farmers and other citizens in the developing world. As biotechnology continues to evolve, factual and open public discourse is vital to define the rote it should play in Society.

Agricultural Biotechnology

1.1 Biotechnology refers to any technique that uses living organisms, or parts of these organisms. Such techniques are used to make or modify products for a practical purpose. Modern medicine, agriculture, and industry make use of biotechnology on a large scale.

1.2 Traditional biotechnologies such as the use of yeast to make bread or wine have been applied for thousands of years. Since the late 19th century, knowledge of the principles of heredity gave farmers new tools for breeding crops and animals. They selected individual organisms with beneficial characteristics and developed hybrid crops.

New methods have been developed since the discovery of the DNA structure in 1954. For instance, micro-organisms can be used to produce antibiotics, and the hereditary material in plants can be changed to make them resistant to pests or diseases. .

2.1 Genes are the pieces of DNA code which regulate all

biological processes in living organisms. The entire set of genetic information of an organism is present in every cell and is called the genome.

2.2 The genetic material is structured in a similar way in different species, which makes it easier to identify potentially useful genes. Certain species of crops, livestock, and disease-causing organisms have been studied as model species because they help us understand related organisms.

2.3 Certain fragments of DNA that can be easily identified are used to 'flag' the position of a particular gene. They can be used to select individual plants or animals carrying beneficial genes and characteristics. Important traits such as fruit yield, wood quality, disease resistance, milk and meat production, or body fat can be traced this way.

2.4 Plants can be obtained from small plant samples grown in test tubes. This is a more sophisticated form of the conventional planting of cuttings from existing plants. Another laboratory technique, in vitro selection, involves growing plant cells under adverse conditions to select resistant cells before growing the full plant.

2.5 In conventional breeding half of an individual's genes come from each parent, whereas in genetic engineering one or several specially selected genes are added to the genetic material. Moreover, conventional plant breeding can only combine closely related plants.

Genetic engineering permits the transfer of genes between organisms that are not normally able to cross breed.

2.6 For example a gene from a bacterium can be inserted into a plant cell to provide resistance to insects. Such a transfer produces organisms referred to as genetically modified (GM) or transgenic.

3. Does conventional plant breeding have effects on health and the environment?

In conventional plant breeding, little attention has been paid to the possible impacts of new plant varieties on food safety or the environment. Nonetheless, this kind of breeding has sometimes caused negative effects on human health. For instance, a cultivated crop variety created by conventional cross breeding can contain excessive levels of naturally occurring toxins.

The introduction of genetically modified plants has raised some concerns that gene transfer could occur in the field between cultivated and wild plants and such concerns also apply to conventional crops. Such transfers have occasionally been reported but are generally not considered a problem.

4. Are genetically modified plant foods safe to eat?

4.1 Foodstuffs made of genetically modified crops that are currently available (mainly maize, soybean, and oilseed rape) have been judged safe to eat, and the methods used to test them have been deemed appropriate. These conclusions represent the consensus of the scientific evidence surveyed by the International Council for Science (ICSU) and are consistent with the views of the World Health Organization (WHO).

However, the lack of evidence of negative effects does not mean that new genetically modified foods are without risk. The possibility of long-term effects from genetically modified plants cannot be excluded and must be examined on a case-by-case basis. New techniques are being developed to address concerns, such as the possibility of the unintended transfer of antibiotic-resistance genes.

Genetic engineering of plants could also offer some direct and indirect health benefits to consumers, for instance by improving nutritional quality or reducing pesticide use.

4.2 Scientists recommend that food safety assessment should take place on a case-by-case basis before genetically modified food is brought to the market. In such assessments, foodstuffs derived from genetically modified plants are compared to their conventional counterparts, which are generally considered safe due to their long history of use. This comparison considers to what extent different foodstuffs can cause harmful effects or allergies and how much nutrients they contain.

4.3 Consumers may wish to select foods on the basis of how they are produced, because of religious, environmental, or health concerns. However, merely indicating whether a product is genetically modified or not, without providing any additional information, says nothing about its content nor about possible risks or benefits. International guidelines are being developed for labelling genetically modified foods.

5. What effects could genetically modified crops have on the environment?

5.1 Agriculture of any type has an impact on the environment. Genetic engineering may accelerate the damaging effects of agriculture, have the same impact as conventional agriculture, or contribute to more sustainable practices.

Growing genetically modified or conventional plants in the field has raised concern for the potential transfer of genes from cultivated species to their wild relatives. However, many food plants are not native to the areas in which they are grown. Locally, they may have no wild relatives to which genes could flow.

Moreover, if gene flow occurs, it is unlikely that the hybrid plants would thrive in the wild, because they would have characteristics that are advantageous in agricultural environments only. In the future, genetically modified plants may be equipped with mechanisms designed to prevent gene flow to other plants.

A controversy has arisen about whether certain genetically modified plants (which are insect resistant because they carry the Bt gene) could harm not only insect pests but also other species such as the monarch butterfly. In the field, no significant adverse effects on non-target species have so far been observed. Nonetheless, continued monitoring for such effects is needed. .

5.2 Genetically modified crops may have indirect environmental effects as a result of changing agricultural or environmental practices. However, it remains controversial whether the net effect of these changes will be positive or negative for the environment. For example, the use of genetically modified insect-resistant Bt crops is reducing the volume and frequency of insecticide use on maize, cotton and soybean. Yet the extensive use of herbicide and insect resistant crops could result in the emergence of resistant weeds and insects.

5.3 The broad consensus is that the environmental effects of genetically modified plants should be evaluated using science-based assessment procedures, considering each crop individually in comparison to its conventional counterparts.

6.1 Animal feeds frequently contain genetically modified crops and enzymes derived from genetically modified micro-organisms. There is general agreement that both modified DNA and proteins are rapidly broken down in the digestive system.

To date no negative effects on animals have been reported. It is extremely unlikely that genes may transfer from plants to disease-causing bacteria through the food chain.

Nevertheless, scientists advise that genes which determine resistance to antibiotics that are critical for treating humans should not be used in genetically modified plants.

6.2 As of 2004, no genetically modified animals were used in commercial agriculture anywhere in the world, but several livestock and aquatic species were being studied. Genetically modified animals could have positive environmental impacts, for example through greater disease resistance and lower antibiotic usage. However, some genetic modifications could lead to more intensive livestock production and thus increased pollution.

7. Are GMOs regulated by international agreements?

7.1 Certain barriers to international agricultural trade have been reduced by the World Trade Organization (WTO). A WTO agreement adopted in 1994 establishes that countries retain their right to ensure that the food, animal, and plant products they import are safe. At the same time it states that countries should not use unnecessarily stringent measures as disguised barriers to trade.

7.2 Several international agreements relate to the environmental aspects of genetically modified crops. The Convention on Biological Diversity is mainly concerned with the conservation and sustainable use of ecosystems but also with environmental effects of GMOs. A part of this convention is the Cartagena Protocol on Biosafety, which regulates the export and import of genetically modified crops.

The International Plant Protection Convention was adopted to prevent the spread of pests affecting plants and plant products. It identified potential pest risks related to GMOs that may need to be considered, such as the potential development of invasive species or effects on beneficial insects and birds.

8. Conclusions

8.1 Agricultural biotechnology can be seen as both:

- a scientific complement to conventional agriculture, aiding for instance plant breeding programs, and
- a dramatic departure from conventional agriculture, enabling transfer of genetic material between organisms that would not normally mix.

Agricultural biotechnology has international implications and may become increasingly important for developing countries.1 However, research has tended to focus on crops important to developed countries.

8.2 To date, countries where genetically modified crops have been introduced in fields, have reported no significant health damage or environmental harm. Moreover, farmers are using less pesticides or using less toxic ones, reducing harm to water supplies and workers' health, and allowing the return of beneficial insects to the fields. Some of the concerns related to gene flow and pest resistance have been addressed by new techniques of genetic engineering.

However, the lack of observed negative effects does not mean that they cannot occur. Scientists call for a cautious case-by-case assessment of each product or process prior to its release in order to address legitimate safety concerns.

"Science cannot declare any technology completely risk free. Genetically engineered crops can reduce some environmental risks associated with conventional agriculture, but will also introduce new challenges that must be addressed. Society will have to decide when and where genetic engineering is safe enough." (FAO 2004)

9. Other views and links

This summary is based on the chapters 2 and 5 of "The

State of Food and Agriculture 2003-2004; Agricultural Biotechnology" of the FAO (Food and Agriculture Organization), a leading scientific report produced by a large international panel of scientists.

It is considered by most scientists as a consensus document and other recent scientific assessments reach similar conclusions, though some people and organizations put forward different views.

Agricultural Biotech: European Attitudes

In the face of staunch public opposition and widespread bureaucratic obstacles at the national and European Union levels, the European agrochemical industry is working hard to create a climate of acceptance for genetically engineered crops. Many agrochemical executives complain that the European Union's Commission (EC), which drafts regulations, tends to overregulate and is suspicious of new technologies.

Indeed, the European Union has fallen behind other parts of the world in research and development of genetically engineered agricultural products. Agrochemical giants such as Germany's Hoescht and Switzerland's Ciba-Geigy have been forced to move to countries such as the United States, which has a pragmatic regulatory scheme, to develop and test their products.

The current backbone of regulatory oversight in Europe for genetically engineered crop plants is the Union's "Directive on the Deliberate Release into the Environment of Genetically Modified Organisms." This regulatory process carries the organisms from the research and development level through field testing and commercialization. Before placing genetically engineered products on the market, manufacturers or importers must demonstrate that such products do not pose risks to human health or the environment.

EU directives are legally binding but require further action by member states concerning implementation and timing. For a transgenic crop plant to complete the journey from field to market in Europe, permission must be obtained first from individual countries for field trials. Depending on the country, this can take years to complete. Next an application must be made to enlarge and prepare the crop for marketing. At this stage, other EU countries can raise objections. Finally, another permit is required to actually market the product.

Recognition is dawning within the EU that if the Union is to avoid becoming simply a market for, rather than a producer of, biotech products, then regulations must be eased. EU officials now are in the process of lightening permit requirements for field tests of organisms and genetic constructs that present low risk to the environment and human health. This basically would replace permits with a notification process and reduce the time periods involved in regulatory procedures. These moves are backed by European industry, which has called for scientifically sound regulations that do not discriminate against products because of the technology used to develop them.

Not all parties in the EU are enthusiastic about the proposed changes. Germany in particular, with its strong Green Party, may be difficult to win over. There have been four instances of vandalism against genetically engineered crops in Germany. France, the United Kingdom, and Belgium are more supportive, although the recent field trial in the UK of a virus modified with a scorpion toxin gene to enhance death rates of target insects on plants has energized opposition to the environmental release of genetically engineered organisms in the UK. In addition, a number of EU member states haven't yet fully implemented the current regulations even though the deadline for implementation was October 1991.

As one Italian official said about the proposal to ease the EU's biotechnology regulations, "We are at the beginning of the discussion on the issue. Our aim is to have the welcoming approach of the U.S."

Agriculture and Biotechnology: Indian Experience

Proponents of free trade argue that liberalization of agriculture will increase food production and improve the economic conditions of farmers around the world. Real life experience, however, is telling a vastly different story as the liberalization of agriculture unfolds. The entry of transnational corporations in the agricultural sector has created devastating conditions in India, leading many small and marginal farmers, for example, to commit suicide due to economic distress.

The entry of transnational corporations in the sector has moved the focus of agriculture from self sufficiency to making profits. On the one hand, reform policies have ended subsidies provided to Indian farmers, leaving them at the mercy of market forces. On the other hand, these policies have increased incentives for large agribusiness companies to farm. Such is the level of disbandment of farmers in India that, as Devinder Sharma, a well known food policy analyst writes:

"In Andhra Pradesh, the state which leads the country in ushering in progressive policies to improve agriculture, chief minister Chandrababu Naidu has launched the Vision 2020 programme that aims to reduce the number of farmers from the present 70 per cent to 40 per cent. While what happens to the remaining 30 per cent of the farming force is none of his concern, the State continues to record the highest suicide rate among the farming community in the country. Such is the government's apathy that instead of resurrecting the policies to help farmers tide over the crisis of livelihood security, chief minister has been talking of deputing psychiatrists to advise farmers not to commit suicide!"

The impacts of liberalization in India are especially important because over 70% of India's population, directly or indirectly, is engaged in the agriculture sector. Add to this the fact the India is home to nearly a third of the world's poor, and one can begin to realize the numbers of people that trade liberalization, especially of the agricultural sector, impacts.

In their pursuit of profits, transnational corporations are eager to push untested technologies and models regardless of the human, environmental and social damages they may cause. India is rapidly moving toward a cash crop economy, at the expense of self-sufficiency. Large multinationals like Monsanto are also pushing the Gene Revolution (biotechnology), as the successor to the Green Revolution, in spite of largely unanswered questions about the impacts of such a technology. There are grand yet unproved claims that biotechnology is the answer to feeding the hungry in the world. Not only are the unanswered questions problematic, biotechnology in agriculture itself is designed so that corporations gain full control over the agricultural system from local farmers.

Ironically, the hunger problem in India is not one of food production but rather food distribution and storage. If left to the free market, as the WTO would like us to do, there will really be no food at all for the poor since we will export all the cash crops and the imported food will be too expensive for the poor to afford!

What is required is food security and achieving food security will involve consciously planning for the needs of the farmers in India, not abandoning them as has been the case since 1991 when the Indian government began to liberalize the economy. It will also involve guaranteeing genuine land reforms for most Indians, especially in the light that land holding size has been steadily declining over the years.

This issue area looks at the problems surrounding liberalization and corporate agriculture in India, and the attempts to turn back the tides of corporate globalization.

Spotlights in Biotechnology and IPR in USA

Strengthening of intellectual property rights (IPR) protection in the 1970s and 1980s increased returns to research and offered greater incentives for private companies to invest in seed development and crop biotechnology. ERS studies IPR protection in agricultural biotechnology and its interrelationship with research activities and industry structure.

(i) The First Decade of Genetically Engineered Crops in the United States: Ten years after the first generation of genetically engineered (GE) crops became commercially available in the U.S. the adoption of GE corn, soybean, and cotton has increased rapidly. Plantings were driven by farmers' expectations of higher yields, savings in management time, and lower pesticide costs. Despite these benefits, environmental and consumer concerns may have limited acceptance of GE crops, particularly in Europe.

(ii) Agricultural Biotechnology Intellectual Property: As the pace of scientific discovery in agricultural biotechnology has accelerated over the past few decades, the use of patents and other intellectual property rights to protect these discoveries has increased. This database identifies and describes U.S. utility patents on inventions in biotechnology and other biological processes used in food and agriculture.

Biotechnology in Food and Agriculture: The FAO Forum Experience

The FAO Electronic Forum on Biotechnology in Food and Agriculture was established in March 2000 with the aim of providing quality balanced information on agricultural biotechnology in developing countries and to make a neutral platform available for people to exchange views and experiences on this subject.

Thirteen moderated e-mail conferences have been hosted by the Forum so far. The first six e-mail conferences, held in 2000-2001, lasted on average 9 weeks. Conferences held from 2002 onwards have been shorter, lasting 4-5 weeks, and 350-650 people have subscribed to each one. In the conferences held so far, over 1100 e-mails have been posted, coming roughly 50:50 from people living in developed and developing countries.

The last conference, Conference 13, ran from 6 June to 3 July 2005 and was entitled "The role of biotechnology for the characterisation and conservation of crop, forest, animal and fishery genetic resources in developing countries". The conference, as usual, was open to everyone and was moderated. About 650 people subscribed and 127 messages were posted, from people in 38 different countries.

The previous conference, Conference 12, ran from 17 January to 13 February 2005 and was entitled Public participation in decision-making regarding GMOs in developing countries: How to effectively involve rural people. Over 500 people subscribed and 116 messages were posted, from people living in 35 different countries.

Conference 11 ran from 14 June to 15 July 2004 and was entitled Biotechnology applications in food processing: Can developing countries benefit? A total of 411 people subscribed to the conference and 68 messages were posted, roughly 70% from people living in developing countries.

Conference 10 ran from 17 November to 14 December 2003 and was entitled Molecular marker assisted selection as a potential tool for genetic improvement of crops, forest trees, livestock and fish in developing countries.

The report of the first six conferences has been published as FAO Research and Technology Paper 8. It is available in English, entitled "Agricultural Biotechnology for Developing Countries - Results of an Electronic Forum". Since November 2003, the report is also available in Spanish, entitled "Biotecnología agrícola para países en desarrollo - Resultados de un foro electrónico", in HTML or PDF format.

Objectives of the Forum

To provide an open forum that will allow a wide range of parties, including governmental and non-governmental organisations, policy makers and the general public, to discuss and exchange views and experiences about specific issues concerning biotechnology in food and agriculture for developing countries. This is done through a series of moderated e-mail conferences, each running for a limited time period only, on specific topics, for which background (before the conferences) and summary documents (after) are produced.

Background to the Forum

Farmers and specialist breeders have developed and used many biotechnologies to improve plants and animals within agriculture, or to make food and agricultural products. Now, improvements in molecular science and in reproductive biology and a radical new understanding of genetics have resulted in the development of a range of new cutting-edge techniques. These allow us to directly modify genetic material, better study the extent/pattern of genetic variation, and they may greatly speed up progress. They may also help us to tackle so far intractable problems. The Forum will focus on such techniques.

Biotechnology is a collection of tools that can be applied to many areas of food and agriculture (including animals, crops, fish and forest trees). This collection comprises scientific tools that are very diverse and sometimes highly controversial. They may pose ethical problems and require substantial debate among policy makers, researchers and the public at large. Particularly in some areas of biotechnology, the debate has become quite polarised and there is therefore an increasing need for quality, unbiased, factual information. It is in this spirit that the Forum was established in March 2000.

FAO Statement on Biotechnology & Agriculture

The statement was published in March 2000 on the occasion of the "Codex Alimentarius Ad Hoc Intergovernmental Task Force on Foods Derived from Biotechnology" meeting in Japan.

Biotechnology provides powerful tools for the sustainable development of agriculture, fisheries and forestry, as well as the food industry. When appropriately integrated with other technologies for the production of food, agricultural products and services, biotechnology can be of significant assistance in meeting the needs of an expanding and increasingly urbanized population in the next millennium.

There is a wide array of "biotechnologies" with different techniques and applications. The Convention on Biological Diversity (CBD) defines biotechnology as:

> *"any technological application that uses biological systems, living organisms, or derivatives thereof, to make or modify products or processes for specific use".*

Interpreted in this broad sense, the definition of biotechnology covers many of the tools and techniques that are commonplace in agriculture and food production. Interpreted in a narrow sense, which considers only the new DNA techniques, molecular biology and reproductive technological applications, the definition covers a range of different technologies such as gene manipulation and gene transfer, DNA typing and cloning of plants and animals.

While there is little controversy about many aspects of biotechnology and its application, genetically modified organisms (GMOs) have become the target of a very intensive and, at times, emotionally charged debate. FAO recognizes that genetic engineering has the potential to help increase production and productivity in agriculture, forestry and fisheries. It could lead to higher yields on marginal lands in countries that today cannot grow enough food to feed their people. There are already examples where genetic engineering is helping to reduce the transmission of human and animal diseases through new vaccines. Rice has been genetically engineered to contain pro-vitamin A (beta carotene) and iron, which could improve the health of many low-income communities.

Other biotechnological methods have led to organisms that improve food quality and consistency, or that clean up oil spills and heavy metals in fragile ecosystems. Tissue culture has produced plants that are increasing crop yields by providing farmers with healthier planting material. Marker-assisted selection and DNA fingerprinting allow a faster and much more targeted development of improved genotypes for all living species. They also provide new research methods which can assist in the conservation and characterization of biodiversity. The new techniques will enable scientists to recognize and target quantitative trait loci and thus increase the efficiency of breeding for some traditionally intractable agronomic problems such as drought resistance and improved root systems.

However, FAO is also aware of the concern about the potential risks posed by certain aspects of biotechnology. These risks fall into two basic categories: the effects on human and animal health and the environmental consequences. Caution must be exercised in order to reduce the risks of transferring toxins from one life form to another, of creating new toxins or of transferring allergenic compounds from one species to another, which could result in unexpected allergic reactions. Risks to the environment include the possibility of outcrossing, which could lead, for example, to the development of more aggressive weeds or wild relatives with increased resistance to diseases or environmental stresses, upsetting the ecosystem balance. Biodiversity may also be lost, as a result of the displacement of traditional cultivars by a small number of genetically modified cultivars, for example.

FAO supports a science-based evaluation system that would objectively determine the benefits and risks of each individual GMO. This calls for a cautious case-by-case approach to address legitimate concerns for the biosafety of each product or process prior to its release. The possible effects on biodiversity, the environment and food safety need to be evaluated, and the extent to which the benefits of the product or process outweigh its risks assessed. The evaluation process should also take into consideration experience gained by national regulatory authorities in clearing such products. Careful monitoring of the post-release effects of these products and processes is also essential to ensure their continued safety to human beings, animals and the environment.

Current investment in biotechnological research tends to be concentrated in the private sector and oriented towards agriculture in higher-income countries where there is purchasing power for its products. In view of the potential contribution of biotechnologies for increasing food supply and overcoming food insecurity and vulnerability, FAO considers that efforts should be made to ensure that developing countries, in general, and resource-poor farmers, in particular, benefit more from biotechnological research, while continuing to have access to a diversity of sources of genetic material. FAO proposes that this need be addressed through increased public funding and dialogue between the public and private sectors.

FAO continues to assist its member countries, particularly developing countries, to reap the benefits derived from the application of biotechnologies in agriculture, forestry and fisheries - through, for example, the network on plant biotechnology for Latin America and the Caribbean (REDBIO), which involves 33 countries. The Organization also assists developing countries to participate more effectively and equitably in international commodities and food trade. FAO provides technical information and assistance, as well as socio-economic and environmental analyses, on major global issues related to new technological developments. Whenever the need arises, FAO acts as an "honest broker" by providing a forum for discussion.

For instance, together with the World Health Organization, FAO provides the secretariat to the Codex Alimentarius Commission which has just established an ad hoc Intergovernmental Task Force on Foods Derived from

Biotechnologies, in which government-designated experts will develop standards, guidelines or recommendations, as appropriate, for foods derived from biotechnologies or traits introduced into foods by biotechnological methods. The Codex Alimentarius Commission is also considering the labelling of foods derived from biotechnologies to allow the consumer to make an informed choice.

Another example is the FAO Commission on Genetic Resources for Food and Agriculture, a permanent intergovernmental forum where countries are developing a Code of Conduct on Biotechnology aimed at maximizing the benefits of modern biotechnologies and minimizing the risks. The Code will be based on scientific considerations and will take into account the environmental, socio-economic and ethical implications of biotechnology. As in applications in medicine, these ethical aspects warrant responsible consideration. Therefore the Organization is working towards the establishment of an international expert committee on ethics in food and agriculture.

FAO is constantly striving to determine the potential benefits and possible risks associated with the application of modern technologies to increase plant and animal productivity and production. However, the responsibility for formulating policies towards these technologies rests with the Member Governments themselves.

Major Reasons for Using Biotechnology in Agriculture

1. Crop production can be increased by identifying and selecting genes resistant to diseases caused by viruses, bacteria and fungi. These genes can then be inserted into other plants.

2. Plants can obtain nutrients from the soil more efficiently reducing the need for fertilizers. *Rhizobium*, a

microbe, can be applied to the seed of plants to assist in nitrogen fixation.

3. Plants can resist insects or better compete with weeds. Researchers in Ontario, Canada, have developed a process whereby a mold on strawberries is controlled using a fungus transferred to plants by honey bees.

4. Farmers can reduce the amount of herbicide and pesticide spraying. Some plants produce chemicals from their roots that retard the growth of other nearby plants. This chemical may assist in the control of weeds.

Techniques aimed at crop improvement have been utilized for centuries. Today, applied plant science has three overall goals: increased crop yield, improved crop quality, and reduced production costs. Biotechnology is proving its value in meeting these goals. Progress has, however, been slower than with medical and other areas of research. Because plants are genetically and physiologically more complex than single-cell organisms such as bacteria and yeasts, the necessary technologies are developing more slowly.

Improvements in Crop Yield and Quality

In one active area of plant research, scientists are exploring ways to use genetic modification to confer desirable characteristics on food crops. Similarly, agronomists are looking for ways to harden plants against adverse environmental conditions such as soil salinity, drought, alkaline earth metals, and anaerobic (lacking air) soil conditions.

Genetic engineering methods to improve fruit and vegetable crop characteristics - such as taste, texture, size, color, acidity or sweetness, and ripening process, are being explored as a potentially superior strategy to the traditional method of cross-breeding.

Research in this area of agricultural biotechnology is complicated by the fact that many of a crop's traits are encoded not by one gene but by many genes working together. Therefore, one must first identify all of the genes that function as a set to express a particular property. This knowledge can then be applied to altering the germlines of commercially important food crops. For example, it will be possible to transfer the genes regulating nutrient content from one variety of tomatoes into a variety that naturally grows to a larger size. Similarly, by modifying the genes that control ripening, agronomists can provide supplies of seasonal fruits and vegetables for extended periods of time.

Biotechnological methods for improving field crops, such as wheat, corn and soybeans, are also being sought, since seeds serve both as a source of nutrition for people and animals and as the material for producing the next plant generation. By increasing the quality and quantity of protein or varying the types in these crops, we can improve their nutritional value. For example, a major protein of corn has very little of two amino acids, lysine and tryptophan, which are essential for human growth. Increased amounts of these amino acids could make corn products a source of improved protein.

Biopesticides and Biofertilizers

Biotechnology makes it possible to develop bacteria essential to herbicide and other pesticide compounds. Certain chemicals produced by these organisms are called allelopathic agents. These chemicals act as natural herbicides, preventing the growth of other plant species in the same geographic area. Black walnut trees, for example, release an allelopathic agent against tomato plants.

Modern high-yield agriculture entails consumption of vast amounts of chemicals for use as fertilizers and as agents to control pests and plant diseases, and any means that will permit the plant to do this work itself could result in significant savings for the farmer. For example, soybeans and certain other legumes produce their own source of usable nitrogen fertilizer by a process known as nitrogen fixation. This process is made possible by a bacterium that grows symbiotically on the plant's roots. (In a symbiotic relationship, dissimilar organisms live together in a mutually beneficial way.) In the nitrogen-fixing process, microbes capture atmospheric nitrogen and biochemically convert it into water-soluble nitrogen. This form of nitrogen is an essential nutrient for increasing the quantity and quality of plant yield.

This bacteria will not assist in the growth of other important crops, such as corn and cereal plants. But research on nitrogen-fixing bacteria and legumes may show how we can modify either the bacteria or non-leguminous plants, thereby making many crops more nearly self-sufficient in obtaining nitrogen. Biotechnology is central to the search for effective, environmentally safe and economically sound alternatives to chemical pesticides. Biotechnology may be used to protect commercial crop plants from insect pests and promises to guard against further environmental deterioration and to provide a useful alternative to traditional methods of insect pest control.

Biopesticides degrade rapidly in the environment - a major environmental benefit. The active elements of bacterial pesticides are proteins that are fragile molecules. Once exposed to the sun and other natural elements, these proteins are quickly broken down, thereby prohibiting spread to groundwater and other animal and plant species. This will help keep our water supply safe to drink, our lakes and streams habitable for water life and recreation.

Unlike chemical pesticide technology, biopesticide technology is based on potent, naturally occurring proteins. These living particles are produced in nature by microorganisms such as Bacillus thuringiensis (B.t.). Discovered at the turn of the century, B.t. has been used without risk in the United States for almost three decades by home gardeners, farmers, and forestry officials. Its active component, a protein, specifically attacks the stomachs of target pests, disrupting their digestive tracks so thoroughly that the pests stop eating and eventually die of starvation. Higher organisms, such as mammals, fish, birds, and other non-target species remain unthreatened, however, because their stomach acid easily breaks down the protein toxin.

The delivery of these biopesticides varies in method and design. In one method, dormant spores of B.t. are dusted on crops. The spores then become active and multiply, covering plants with a bacteria poisonous to the target insects that feed on them. The B.t. toxin gene can also be inserted into the genetic makeup of crops, giving them a built-in resistance to insects. Similarly, the toxin gene can be put into a third party, such as a microorganism that lives within the plant's sap. These organisms - known as endophytes - multiply within the host plant and move throughout the plant's vascular system, forming a microscopic defense against feeding insects. This process resembles vaccines moving throughout a person's vascular system to defend against harmful disease.

Some of the concerns farmers raise about having to use increasingly dangerous pesticides to produce adequate crops may well be addressed by biotechnology. Further research in agricultural biotechnology and biopesticide development aims to provide attractive alternatives to the farmer that will lower overall unit cost of production and allow the farmer to be more competitive in the highly cost-sensitive world markets. While some uses of chemical pesticides will be necessary for decades to come, continued development by biotechnology companies of useful biological pesticides will offer farmers viable alternatives.

Biotechnology in Agriculture: A Review

Nineteen ninety-four was a turbulent year for the biotechnology industry. The industry's largest sector, biopharmaceuticals, was hit with a number of clinical trial setbacks and failures, and investors began to question the worthiness of their high risk investments in biotechnology in general (1). In fact, three of the top ten worst performing stocks for 1994 on the NASDAQ market (based on percentage change from 1993) were biotechnology stocks: Telios Pharmaceuticals, MicroProbe, and Alpha 1 Biomedicals (2). Although each of these three companies are focused on therapeutic development, agricultural biotechnology companies may have been hurt by the pharmaceutical sector, and did not fare much better.

Here is a sample of some select ag-biotechnology company stocks and their percentage price change during 1994 (3):

- Crop Genetics International (-82%)
- AgriDyne Technologies (-71%)
- Calgene (-43%)
- Ecogen (-43%)
- DNA Plant Technology (-36%)
- Mycogen (-18%)

As was noted in the Agricultural Biotechnology Notes section from the August 1994 NBIAP News Report, the movement of agricultural biotechnology stocks closely mimics that of the biopharmaceutical sector. This is likely due to the large swings in individual biotechnology company stock prices (usually therapeutic firms) that pull the entire biotechnology industry with them, economic conditions that impact all stocks, and a lack of sophistication and understanding of the technologies by some investors.

In addition to the poor market showing by agricultural biotechnology companies, a number of other notable developments took place in 1994 related to the agricultural biotechnology sector. Some of these include:

- Monsanto's bovine somatotropin (BST) went on the market in February 1994, creating controversy and litigation over its economic impact and product labeling provisions.
- Calgene's genetically engineered Flavr Savr (TM) tomato was approved by the FDA in April 1994, and went on the market in the United States. It was also issued patent protection by the European Union.
- The USDA (APHIS) approved discontinuation of regulation under the Plant Pest Act of some genetically engineered agricultural products, including BXN (TM) cotton (a herbicide-tolerant cotton), Laurate Canola, the Flavr Savr (TM) tomato (all developed by Calgene), and Monsanto's herbicide-tolerant soybean.
- The NIH published revised guidelines for research involving recombinant DNA molecules.
- The FDA developed proposed regulations that would establish a mandatory premarket notification system for foods derived from genetically modified plants.
- The German government eased its strict biotechnology regulations, making it easier to conduct research and development using genetic engineering.

One area that this change may impact is the number of field trials conducted in Germany, as only 2 field release permits had been approved there between 1986 and 1992, compared with 77 in France and 316 in the United States for that same period. The European Union also worked toward easing permit requirements for field tests of genetically engineered plants.

- The broad patent awarded to Agracetus in 1992, covering all genetically engineered cotton varieties, came under scrutiny and was considered for possible re-examination.
- The U.S. Patent Office requested comments on proposed changes in patent protection rules for biological inventions.

The FDA found seven genetically engineered food plants safe, including tomatoes from Zeneca Plant Science, DNAP Plant Technology, and Monsanto; ZW-20 squash from Asgrow; insect-resistant potato and herbicide-tolerant soybean from Monsanto; and herbicide-tolerant cotton from Calgene. Although these food crops were deemed safe by the FDA, they are also subject to approval by the USDA and/or the EPA.

In spite of the lack of confidence of investors in agricultural biotechnology companies during 1994, one end-of-the-year forecast predicts almost 30% average annual sales growth over the next ten years of ag-biotech products, increasing from an estimated $130 million in 1994 to almost $1.5 billion in 2004 (4). Research discoveries continue to evolve, and this, combined with breakthroughs in ag-product commercialization and regulation in 1994, continue to fuel optimism for the sector's future growth.

Myriad Genetics and Syngenta Complete Rice Genome Map

January 26, 2001 at Salt Lake City

Myriad Genetics, Inc. (Nasdaq: MYGN) announced on that it has completed DNA sequencing of the entire rice genome in collaboration with the Torrey Mesa Research Institute, TMRI, (formerly Novartis Agricultural Discovery Institute, Inc.), a subsidiary of Syngenta. Myriad completed the high-throughput sequencing component of the genome six months ahead of schedule and under budget, triggering a $3 million cash bonus payment to Myriad.

The rice genome map will assist crop researchers and plant breeders in the development of new crop varieties with enhanced characteristics. Additionally, it may serve as a model for other cereal crops such as corn and wheat. In addition to the DNA sequence, the information about the rice genome includes the regulatory DNA sequences controlling gene activity and the location of most of the genes. Furthermore, the collaboration will now be expanded to include the use of Myriad's ProNetÂ® technology to elucidate the rice proteome. The sequencing of the rice genome is an important milestone because it is the first agriculturally important crop to be sequenced. The rice genome is made up of 430 million bases of DNA, representing approximately 50,000 genes, making it the second largest genome of any kind to have been sequenced to date.

Myriad and Syngenta completed the rice genome to a very high degree of accuracy in less than a year and a half from announcement of the project. The rice genome sequence is the most accurate and comprehensive large genome with direct commercial value yet sequenced. The rice genome map is better than 99.5% complete, compared with the human genome map, which still contains a number of large gaps and is estimated to be approximately 90 - 95% complete.

"We are very proud that the Torrey Mesa Research Institute and Myriad have pioneered this research and that our innovative approach has been successful," said Dr. Steve Briggs, President of TMRI. "Identifying not only the genes, but their functions and how they work, will provide researchers with crucial new knowledge to improve food crops."

This remarkable accomplishment was aided by Myriad's proprietary, high-throughput DNA sequencing facility that averages 40% more sequence per machine than other users of capillary sequencing machines. Myriad's outstanding capability in DNA sequencing pervades the organization, from predictive medicine to proteomics and genome sequencing. The expansion of the collaboration adds the use of the Company's ProNetÂ® technology to determine how protein interactions control the rice plant's growth and development.

"Our Syngenta collaboration highlights many of Myriad's strengths, including our strategy of building multiple sources of near-term revenue generation," said Peter Meldrum, President and Chief Executive Officer of Myriad Genetics, Inc.Â "The alliance has demonstrated our ability to move away from mere royalties toward 50:50 profit sharing agreements on the use of our technologies."

The rice genome project required extensive use of Myriad's genomic technologies, including its high-throughput DNA sequencing facility that is based upon the latest generation of capillary DNA sequencers. The facility is generating over 12 million bases of DNA sequence daily, a rate that would allow sequencing the entire human genome in a single year. The high throughput DNA sequencing facility is used in most of Myriad's research programs for the delivery of high quality, high throughput DNA sequence.

"Myriad's bioinformatics capabilities made this accomplishment possible in such a short time frame," said Sudhir Sahasrabudhe, Ph.D., Executive Vice President of Research and Development for Myriad Genetics, Inc.Â "The basecalling and sequence assembly software as well as software that constructs individual chromosomes from the segmented data were all developed in-house at Myriad and are available only to Myriad and its collaborators."

Myriad and Syngenta will make the rice genome data available to researchers worldwide through research contracts. In the developing world, where rice is a vital crop, Syngenta will work with local research institutes to explore how this information can best be used to find crop improvements to benefit subsistence farmers.Â It is Syngentaâ•™s policy to provide such information and technology for subsistence farm-related research without royalties or technology fees. Rice is the largest commercial crop in the world, nourishing over half of the worldâ•™s population. Knowledge of the rice plant's genes and their function may help plant scientists to introduce new varieties that have enhanced resistance to pests and disease. Additionally, higher yielding varieties could make a major contribution toward the elimination of world hunger.

To access a live Webcast to discuss this scientific breakthrough, please log onto http://www.syngenta.com/, and click on the teleconference icon, at 10:45am eastern time today, for a presentation hosted by Myriad Genetics and Syngenta.

Syngenta is the world's leading agribusiness. The company ranks first in crop protection, and third in the high-value commercial seeds market. Proforma sales in 1999 were approximately US $7 billion. Syngenta employs more than 20,000 people in over 50 countries. The company is committed to sustainable agriculture through innovative Research and Technology. Formed in November 2000 by the merger of Novartis Agribusiness and Zeneca Agrochemicals, Syngenta is listed on the Swiss stock exchange, and in London, New York and Stockholm. Further information is available at http://www.syngenta.com/.

Myriad Genetics, Inc. is a biopharmaceutical company focused on the development of novel therapeutic products derived from its proprietary genomic and proteomic technologies. The Company has established two wholly owned subsidiaries. Myriad Pharmaceuticals, Inc. develops and intends to market therapeutic compounds, and Myriad Genetic Laboratories, Inc. develops and markets proprietary predictive medicine and personalized medicine products. The Company has established strategic alliances with Bayer, Eli Lilly, Hitachi, Novartis, Pharmacia, Roche, Schering AG, Schering-Plough and Syngenta.

The discussion in this news release includes forward-looking statements that are subject to certain risks and uncertainties, including statements relating to the availability of the DNA sequence data to third parties, the use of the rice genome as a model for other cereal crops, the expansion of the collaboration to include ProNetÂ® technology, the knowledge that researchers will gain from the data, the benefit to Myriadâ•™s pharmacogenomics effort from the additional high-throughput DNA sequencing capacity, the accomplishment of plant scientists based on knowledge of the rice plantâ•™s genes and function, the contribution toward the elimination of world hunger that higher yielding varieties could make, and Myriad Pharmaceuticals, Incâ•™s intent to market therapetic compounds. Such statements are based on managementâ•™s current expectations that are subject to risks and uncertainties that could cause actual results to differ materially from those set forth or implied by forward-looking statements, including, but not limited to, intense competition related to the discovery of disease-related genes; uncertainties as to the extent of future government regulation of the Company's business, uncertainties as to whether the Company and its collaborators will be successful in developing, and obtaining regulatory approval for, and commercial acceptance of, therapeutics based on the discovery of disease-related genes and proteins; uncertainties as to the Company's ability to develop therapeutic lead compounds, which is a new business area for the Company; the risk that markets will not exist for therapeutic lead compounds that the Company develops or if such markets exist, that the Company will not be able to sell compounds, which it develops, at acceptable prices.

2

Biotechnology, Farmers and Biotech Crops

U.S. farmers are widely adopting biotech crops, USDA Reports. Genetically engineered soyabeans, cotton, corn are most rapidly accepted U.S. farmers have adopted genetically engineered (GE) crops widely since their introduction in 1996, notwithstanding uncertainty about consumer acceptance and economic and environmental impacts. Soybeans and cotton genetically engineered with herbicide-tolerant traits have been the most widely and rapidly adopted GE crops in the U.S., followed by insect-resistant cotton and corn.

Extent of Adoption

Herbicide-tolerant (HT) crops, developed to survive application of specific herbicides that previously would have destroyed the crop along with the targeted weeds, provide farmers with a broader variety of options for effective weed control. Based on USDA survey data, HT soybeans went from 17 percent of U.S. soybean acreage in 1997 to 68 percent in 2001 and 89 percent in 2006. Plantings of HT cotton expanded from 10 percent of U.S. acreage in 1997 to 56 percent in 2001 and 65 percent in 2006. The adoption of HT corn has lagged soybeans and cotton, reaching 36 percent of U.S. corn acreage in 2006.

Insect-resistant crops containing the gene from the soil bacterium Bt (Bacillus thuringiensis) have been available for corn and cotton since 1996. These bacteria produce a protein that is toxic to specific insects, protecting the plant over its entire life. Plantings of Bt corn grew from 8 percent of U.S. corn acreage in 1997 to 26 percent in 1999, then fell to 19 percent in 2000 and 2001, before climbing to 29 percent in 2003 and 40 percent in 2006. The recent increases in acreage share may be largely due to the commercial introduction in 2003-2004 of a new Bt corn variety that is resistant to the corn rootworm, a pest that may be more destructive to corn yield than the European corn borer, which was previously the only pest targeted by Bt corn. Plantings of Bt cotton expanded more rapidly, from 15 percent of U.S. cotton acreage in 1997 to 37 percent in 2001 and 57 percent in 2006.

Use of Bt corn will likely continue to fluctuate over time, based on expected infestation levels of European corn borer (ECB), and the corn rootworm, which are the main pests targeted by the Bt corn. Similarly, adoption of Bt cotton depends on the expected infestation of Bt target pests, such as the tobacco budworm, the bollworm, and the pink bollworm. Adoption appears to have reached the low-growth phase, as adoption has already occurred on acreage where Bt protection is needed most. Insects have not posed major problems for soybeans, so insect-resistant varieties have not been developed.

These figures include adoption of "stacked" varieties of cotton and corn, which have both HT and Bt traits. Stacked cotton reached 39 percent of cotton plantings in 2006. Plantings of stacked corn are lower, making up only 15 percent of corn acres in 2006.

Adoption of all GE cotton, taking into account the acreage with either or both HT and Bt traits, reached 83 percent in 2006, versus 89 percent for soybeans. In contrast, adoption of all biotech corn was 61 percent.

Genetic Engineering is the introduction of a specific gene into the DNA of a plant to obtain a desired trait. The gene introduced may come not only from another plant species, but also from other organisms. While traditional plant breeding involves crossing related plants, biotechnology is a new tool that enhances the capability of breeders to be more precise.

The goals of genetic engineering are the same as with traditional breeding. They may aim to improve crop performance in the field by conferring pest and disease resistance, herbicide resistance, or tolerance to environmental stresses (such as drought or flooding). They may also aim to develop products with enhanced value, such as improved post-harvest life, nutritional value, or other health benefits.

Insect resistance

In the last few years, several crops have been genetically engineered to produce their own Bt proteins, making them resistant to specific groups of insects. "Bt" is short for Bacillus thuringiensis, a soil bacterium that contains a protein that is toxic to a narrow range of insects, but not harmful to animals or humans. Applications of Bt bacteria have been used to control insect pests for many years, before the advent of the current Bt crops made using biotechnology.

Varieties of Bt insect-resistant corn and cotton are now in commercial production. Other crops being investigated include cowpeas, sunflower, soybeans, tomatoes, tobacco, walnut, sugar cane, and rice.

Herbicide tolerance

Chemical herbicides are frequently used to control weeds. Weeds growing in the same field with crop plants can significantly reduce crop yields because the weeds compete for soil nutrients, water, and sunlight. Many farmers now control weeds by spraying herbicides directly onto the trop plants. Because these herbicides generally kill only a narrow spectrum of plants (if they didn't, they would kill the crop plants, too), farmers apply mixtures of multiple herbicides to control weeds after the crop bas started to grow.

Researchers realized that if a crop plant is genetically engineered to be resistant to a broad-spectrum herbicide, weed management could be simplified and safer chemicals could be used. It is often argued that such GE varieties reduce soil erosion, because they make adoption of soil-conserving practices such as "no-till" easier. Resistance to synthetic herbicides bas been genetically engineered into corn, soybeans, cotton, canola, sugar beets, rice, and flax. Some of these varieties are commercialized in several countries. Research is ongoing on many other crops. One application of this technology is that herbicide could be coated on seed from an herbicide resistant variety (for example, maize) and while the maize would germinate and thrive, weeds and parasites such as Striga would be killed.

Virus resistance

Many plants are susceptible to diseases caused by viruses, which are often spread by insects (such as aphids) from plant to plant across a field. The spread of viral diseases can be very difficult to control and crop damage can be severe. Insecticides ore sometimes applied to control populations of transmitting insects, but often have little impact on the spread of the disease. Often the most effective methods against viral diseases are cultural controls (such as removing diseased plants) or plant varieties bred to be resistant (or tolerant) to the virus, but such strategies may not always be practical or available. Scientists have discovered new genetic engineering methods that provide resistance to viral disease where options were limited before.

- In the U.S., several varieties of squash and zucchini resistant to three important viral diseases have been developed and commercialized.
- Beginning in 1992, a devastating outbreak of Papaya Ring Spot Virus (PRSV) swept through the papaya plantations of Hawaii — papaya production dropped 40 percent in the course of 5 years. Researchers in Hawaii and at Cornell University developed two GE varieties of papaya resistant to PRSV Papaya growers in Hawaii have been able to grow GE virus resistant papaya since 1998.
- Scientists are currently developing virus-resistant crops for Africa, including cassava, maize and sweet potato.

Delayed fruit ripening

Delaying the ripening process in fruit is of interest to producers because it allows more time for shipment of fruit from the farmer's fields to the grocer's shelf, and increases the shelf life of the fruit for consumers. Fruit that is genetically engineered to delay ripening can be left to mature on the plant longer, will have longer shelf-life in shipping, and may last longer for consumers.

Foods with improved nutritional value

Researchers are using biotechnology for the development of foods with improved nutritional value. Genetic modification can be used to produce crops that contain higher amounts of vitamins to improve their nutritional quality. Genetically altered "golden rice," for example, contains three transplanted genes that allow plants to produce beta-carotene, a compound that is converted to vitamin A within the human body. Vitamin A deficiency — the world's leading cause of blindness — affects as many as 250 million children. Biotechnology has also been used to alter the content of many oil crops, either to increase the amount of oil or to alter the types of oils they produce. Biotechnology could also be used to upgrade some plant proteins now considered incomplete or of low biological value because they lack one or more of the 'essential' amino acids. Examples include maize with improved protein balance and sweet potatoes with increased total protein content. Reducing toxicity of certain foods is also a goal of biotechnology. For example, reduction of the toxic cyanogens in cassava has been shown to be possible and could be produced in the future.

Assessing Biotech Crops for Safety

Environmental safety includes in assessments, the USAID agency says.

Commercially available foods and crops made using biotechnology have been subjected to more testing and regulation than any other agricultural products, and have all been found safe. This note is an introduction to the assessment processes that have been used in determining the food and environmental safety of these products.

Food Safety Assessment

Although biotechnology broadens the scope of genetic changes that can be introduced into plants used for food, it does not inherently result in foods that are less safe than those produced by other techniques. This means that previously established principles for assessing food safety still apply for products of biotechnology. Moreover, these products can be judged on their individual safety, allergenicity, toxicity and nutrition rather than their method of production.

The safety assessment of biotech foods is based on the principle that they can be compared with traditional foods that have an established history of safe use. This comparison can be based on an examination of the same risks that have been established for traditional foods, keeping in mind that many conventional foods may present specific risks depending on conditions of processing, or to individuals within a population. The intent of this comparative approach is to establish whether the new food is "as safe as" as" its conventional counterpart.

International discussion and expert consultations have resulted in a consensus on the specific safety issues that should be considered when evaluating these new foods. They include the following:

- The Parent Plant: Knowledge of the biology of the plant and its history of safe use as a food are important for understanding the natural range and variation of key nutritional components, such as vitamins and minerals; naturally occurring toxicants; anti-nutrients; and potential allergens.
- The Gene Source: Information about the natural history of the source organism of a new gene is important in assessing whether it could be a potential source of allergens or toxins.
- Nutrition: All plant breeding methods have the potential to change the nutritional value of plants or to lead to unexpected changes in levels of natural toxicants or anti-nutrients. Food safety assessments take into account the potential for any change in nutritional composition, especially in key components that have a significant impact on the diet. Laboratory

analyses are used to compare the profile of the new plant with its conventional counterpart for constituents such as protein, fat, fibre, micronutrients, amino acids, fatty acids, vitamins, toxins and anti-nutrients.

- Allergens: The potential of accidentally introducing a new allergen (a protein that causes an allergic reaction) into a food is an important safety concern. Fortunately, food allergens have common characteristics, such as remaining stable during digestion and food processing, and they are usually abundant in foods. Internationally accepted approaches for evaluating allergenicity have been established and none of the new proteins present in biotech foods share the properties of allergens.
- Toxins: The possibility that new toxins may have been introduced into a food is also tested by comparing any newly produced proteins with known protein toxins using laboratory analyses and animal studies. Like allergens, protein toxins have very well characterized properties, which are not shared by any of the new proteins in biotech foods.

In assessing these safety issues, the processed version of the food and its impact on specific population subgroups, such as infants or the elderly, is also evaluated.

Environmental Safety

Assessing the environmental safety of a biotech plant requires an understanding of the biology of the plant itself and the practices used in its cultivation. This knowledge is important in identifying and evaluating potential environmental rises and also in designing any appropriate risk management measures. Most countries use similar environmental risk assessment approaches, which include:

- Evaluating the role of the introduced gene in the plant and any changes in the plant's characteristics
- The possible unintended secondary effects on non-target organisms
- The possibility that the modified plant could persist longer in the environment or invade new habitats
- The likelihood and consequences of the potential spread of newly introduced traits to related plants
- Potential impacts on biodiversity

These environmental safety concerns are not unique to plants produced using biotechnology and are also important when evaluating new varieties produced through conventional plant breeding. The objective of environmental safety assessment is to identify and evaluate any additional risks associated with the release and cultivation of these new plants in comparison with a conventional crop variety that has a history of safe use.

In addition to considering the potential risks associated with the introduction of new biotech crop varieties, consideration must be given to the risks associated with not using biotechnology to achieve desired goals. For example, the biodiversity of tropical rainforests or other ecologically sensitive areas can only be maintained if these natural ecosystems are not destroyed because of the expansion of the agricultural land base. Biotech crops can alleviate pressure to expand agricultural areas by increasing yields wits improved best resistance and increased tolerance to drought or saline soil conditions.

Establishing National Biosafety Systems

Developing a Biosafety System

All countries with functioning biosafety regulatory systems have developed these systems gradually, usually beginning with voluntary guidelines and standards developed cooperatively by stakeholders in academia, industry, and government. Over time these were incorporated in laws, either under existing legislation covering food and agricultural products or new legislation dealing specifically with biotechnology. Even in countries with long-established systems, biosafety policy and its implementation continue to evolve and it is not unusual to have a mix of voluntary and mandatory measures.

For countries seeking to develop a national biosafety system, it must be emphasized that there is no model for a single best approach. The issues to be considered can be broadly divided into six elements, which are briefly discussed below.

National Inventory and Evaluation

An inventory and evaluation of national priorities, agricultural policies, existing regulatory regimes, and national scientific and technical capacities, is an ideal prerequisite to the development and implementation of biosafety-related policies and regulations. This national appraisal provides a means to identify and characterize available resources and regulatory infrastructures, assess their adequacy for supporting a biosafety system, and identify gaps where capacities need to be strengthened.

National Policies and Strategies

A national biosafety policy or strategy provides a set of principles to guide the development and implementation of a biosafety system and should describe the goals and objectives of the regulatory framework. Direction on many of the fundamental issues and public policy choices that must be considered during the development of regulations can be provided by such a strategy. Examples of these issues include the extent to which social, ethical, and economic factors should be considered, the social acceptability of biotechnology and its products, and linkages with other national policies on food, agriculture, and economic development.

Scientific Knowledge, Skills and Capacity Base

The human resource environment that both enables and limits biosafety implementation is shaped by the scope and quality of: competency in the biological sciences; expertise in information acquisition, communications, and management; and, experience in critical thinking, analysis, and decision-making. These capacities have an overriding influence on the development and implementation of a biosafety system. Addressing capacity needs is the top priority for many developing countries.

Building a strong base of scientific knowledge in support of the regulatory system, and development of core competencies in biotechnology product evaluation, are fundamental to any national biosafety system. These activities allow an improved scientific basis for assessments of potential risks and/or benefits, and they strengthen the scientific capabilities for risk management, inspection, and monitoring.

Development of Regulations

Decisions on an appropriate regulatory framework should be informed by the national inventory and evaluation, and through extensive consultation with stakeholders, including the public. This is particularly true if a country chooses to incorporate non-safety issues into its decision-making process.

Implementation of Regulations

The central issues around the implementation of biosafety regulations involve the establishment of appropriate mechanisms for risk assessment, risk management, and risk communication within existing financial, technical, and human resource constraints. Decisions made during the implementation phase directly affect the costs associated with assessing and managing risks and ensuring compliance with regulations.

Cross Cutting Issues

Cross cutting issues are those that are common to each of the five preceding elements and they are often the most challenging factors to address and resolve. They are, however, the issues that will ultimately dictate the scope of a national policy on biosafety, and the conversion of policy into practice. Cross cutting issues affect the implementation of the system designed to assess biosafety, and perhaps more importantly, those non-technical factors that are crucial to public acceptance and confidence in the decisions that are made by government on behalf of the people.

The twin issues of public information and participation have to do with the degree of transparency in a regulatory system, and the degree to which the public has input either into the formulation of regulatory policy or into specific regulatory decisions. Transparency refers to the extent to which governments provide information on why and how certain products are regulated, how risk assessments are performed and decisions made, and as well, the conclusions and decisions that have been reached. Transparency can also involve the perceived independence and objectivity of the regulatory decision-makers.

Human, financial and infrastructure resources largely determine the scientific and administrative capacity of any country; they obviously influence any biosafety related policy or program. Funds must be available to develop and implement a national biosafety system; to support the infrastructure required, such as buildings, labs, equipment, and computers; to facilitate communication and public participation; to train scientific and regulatory personnel; and to foster the research required to assure that risk assessments are sound.

The development of an effective national biosafety system is important to encourage the growth of domestic biotechnologies; to ensure safe access to nevi products and technologies developed elsewhere; and to build public confidence that products in the marketplace are safe. The absence of a suitable framework affects the ability of the public and private sectors to invest in biotechnology and to make the products of biotechnology available so that the benefits they afford can be realized.

Biotechnology and U.S. Agricultural Trade

Agricultural biotechnology is a collection of scientific techniques, including genetic engineering, that are used to create, improve, or modify plants, animals, and microorganisms. Using conventional techniques, such as selective breeding, scientists have been working to improve plants and animals for human benefit for hundreds of years. Modern techniques now enable scientists to move genes (and therefore desirable traits) in ways they could not before—and with greater ease and precision.

Combating human diseases—The first biotechnology products were medicines designed to address human diseases. Insulin, used to treat diabetics, and blood clot-busting enzymes for heart attack victims are now produced easily and cheaply as a result of biotechnology. New research by USDA's Agricultural Research Service (ARS) finds that transgenic animals may be a new source of valuable hormones and drugs to treat emphysema and infections in babies.

Promoting human health—Researchers are creating ways to boost the nutritional value of foods using biotechnology. Combating animal diseases—Biotechnology helped produce a vaccine that protects animals in the wild against rabies and a vaccine for "shipping fever" of cattle, the biggest killer of beef cattle in feedlots. Fighting hunger by resisting plant diseases and increasing crop yields—Biotechnology can help farmers increase crop yields and feed even more people. For example, an ARS scientist used biotechnology to pinpoint a gene that could help wheat, a major food staple, grow on millions of acres worldwide that are now hostile to the crop. ARS scientists have also developed an experimental potato hybrid that contains genes to resist a new, more virulent strain of the so-called "late blight," the disease that caused the Irish potato famine in the 1840s.Helping the environment by reducing pesticide use—Biotechnology can help farmers reduce their reliance on insecticides and herbicides. For example, Bt cotton, a widely grown biotech crop, kills several important cotton pests.

Some people may not feel comfortable with new technology. They may worry about environmental, social and economic consequences—real or simply feared. Particularly in Europe, these consumer concerns have led to a resistance toward biotechnology-derived plants and food, though not medicine. European consumer concerns may also be bolstered by a historic distrust in their governments' ability to ensure the safety of the food supply after the "mad cow" scare and more recent dioxin contamination. Consequently, European governments have been slow to approve and permit the importation of biotechnology-derived products, even if they have been proven safe to human and animal health and the environment. This has created friction with the United States and other trading partners, who believe sound science should be the basis of regulatory systems. The failure of other countries to develop consistent and science-based regulatory processes governing biotechnology has the potential to constrain scientific innovation and disrupt trade. USDA recognizes and appreciates differences in the political, social, and cultural situations in countries around the world. However, regulatory processes must be predictable and transparent if they are to be understood and trusted by industry and the public.

The Federal government has a well-coordinated system to ensure that new agricultural biotechnology products are safe for the environment and to animal and human health. While these agencies act independently, they have a close working relationship.

USDA's Animal and Plant Health Inspection Service (APHIS) is responsible for protecting American agriculture against pests and diseases. The agency regulates the field testing of genetically engineered plants and certain microorganisms. APHIS also approves and licenses veterinary biological substances, including animal vaccines, that may be the product of biotechnology.

USDA's Food Safety and Inspection Service (FSIS) ensures the safety of meat and poultry consumed as food. The Department of Health and Human Service's Food and Drug Administration (FDA) governs the safety and labeling of drugs and the nation's food and feed supply, excluding meat and poultry. The Environmental Protection Agency (EPA) ensures the safety and safe use of pesticidal and herbicidal substances in the environment and for certain industrial uses of microbes in the environment. The Department of Health and Human Service's National Institutes of Health have developed guidelines for the laboratory use of genetically engineered organisms. While these guidelines are generally voluntary, they are mandatory for any research conducted under Federal grants and they are widely followed by academic and industrial scientists around the world.

APHIS is the government's lead agency regulating the safe testing, under controlled circumstances, of biotechnology-derived, new plant varieties. A company, academic or research institution, non-profit organization or public sector scientist wishing to field test or move (via importation or interstate movement) a biotechnology-derived plant must generally obtain APHIS approval before proceeding. If someone wants to test a biotechnology-derived plant,what do they do?Applicants ask APHIS for permission to allow field testing (environmental release). They provide information about the plant, including all new genes and gene products, their origin, the purpose of the test, how the test will be conducted, and specific precautions to be taken to prevent the escape of pollen, plants, or plant parts from the field test site. An APHIS scientific reviewer evaluates the possible environmental impacts of the proposed field test. The possible impact of new plant varieties on endangered or threatened species is considered. Non-target species, those not meant to be directly impacted by the new plant, are also taken into account.In an effort to streamline the permit process, most applicants can now use a simplified procedure in which they notify APHIS before they plan to move or field test a biotech crop. APHIS then has 30 days to review the notification prior to any testing. These field tests are still required to meet all the same safety standards as trials that are approved through the permit process.

If testing is approved, APHIS personnel and state agriculture officials may inspect the field test site before, during, or after the test to ensure that is conducted and managed safely.

Generally, before a genetically engineered crop can be produced on a wider scale and sold commercially, its creators must petition APHIS for a "determination of non-regulated status," which requires the submission of more information than a field test permit request. APHIS must be provided scientific details about the genetics of the plant, the nature and origin of the genetic material used, information about indirect effects on other plants, field test reports, and even information unfavorable to the petition. All petitions are published in the Federal Register and the public is given time to comment. APHIS grants the petition only if it determines that the plant poses no significant risk to other plants in the environment and is as safe to use as more traditional varieties.

This is a determination by APHIS that the new plant should be treated like any other plant and, therefore, may be grown, tested, or used for traditional crop breeding without any additional APHIS action. Essentially, this determination permits the plant to be widely grown and commercialized.

Authorizations are also required for the movement into the United States or between states of any genetically engineered organism that is a potential plant pest. And, some developers of the very few biotech plants that are not regulated by APHIS may seek a voluntary courtesy permit, which may make it easier to move or field test the plant.

With APHIS approval, over 5,000 field trials have been safely conducted since 1987. About 40 new agricultural products have completed all the federal regulatory requirements (from all relevant agencies) and may be sold commercially. They range from longer-lasting tomatoes to pest-resistant corn.USDA's National Agricultural Statistics Service (NASS) maintains statistics for all crops in the United States, including biotech crops. Statistics for crops can be found at www.usda.gov/nass. As of June 2002, NASS data indicate that of all U.S. corn, cotton, and soybeans planted, 34 percent, 71 percent, and 75 percent, respectively, are biotechnology varieties.A report from the International Service for the Acquisition of Agri-biotech Applications (ISAAA) estimates total acres of transgenic crops by country. According to the ISAAA, in 2001, the United States grew an estimated 88 million acres of transgenic crops. Estimated globally, the number of acres of transgenic crops in 2001 was 130 million acres.

EPA approves new herbicidal and pesticidal substances. EPA also issues permits for large scale testing of herbicides and biotechnology-derived plants containing new pesticidal substances. In deciding whether to register a new pesticide, EPA considers human safety, the fate of the substance in the environment, the safety for humans, its effectiveness on the target pest, and any effects on other, so-called "non-target" species.

FDA ensures that foods derived from new plant varieties are safe to eat, holding them to the same high standard of safety as any more traditional food product.

The commercialization of Monsanto's genetically engineered Bt Cotton in India is a done deal. Despite a controversy-ridden evaluation process, the transgenic plant has been approved with no data made public about its performance and potential side effects. Critics of the Indian Government's handling of the Bt Cotton evaluations say that the Government has condoned serious irregularities and safety violations by Monsanto-Mahyco. They complain that the process lacks transparency and public debate; and that the Government has neither the political will nor the technical and infrastructural ability to monitor or regulate this controversial technology.

Bt Cotton in India: A Chronology

March 10, 1995: Department of Biotechnology (DBT) of the Government of India permits import of 100 gm of transgenic Cocker-312 variety of cottonseed cultivated in the United States by Mahyco. This variety contained the Cry 1 Ac gene from the bacterium Bacillus thuringiensis.

April 1998: Monsanto-Mahyco tie up. Monsanto given permission for small trials of Bt cotton 100 g per trial by Department of Biotechnology (DBT).

November 28, 1998: Thousands of farmers occupied and burned down Bt cotton trial fields in Karnataka as part of Operation Cremation Monsanto.

January 6, 1999: Vandana Shiva's Research Foundation for Science, Technology and Ecology goes to the Supreme Court challenging the "illegality" of the field trials authorized by the DBT.

July 2000: Mahyco allowed to conduct large-scale field trials including seed production at 40 sites in six states. The permission was granted based on the "totally confidential" data from the small trials that allowed regulators to infer that Bt Cotton was "safe."

The DBT sets up Committee to "independently" monitor evaluate large scale field trials.

January 5, 2001: A 10-member delegation comprising U.S. judges and scientists meet Supreme Court Chief Justice A.S. Anand to educate him — and other members of the judiciary — on biotechnology.

June 18, 2001: An open dialogue held between Monsanto and Greenpeace to discuss Bt cotton with scientists, Ministry of Environment representatives and farmers. No data on field trials was presented, though farmers vociferously demanded Bt cotton be commercialized. Technical questions and concerns raised by Greenpeace remain unanswered.

June 19, 2001: Genetic Engineering Approval Committee (GEAC) extends field trials of Bt Cotton by another year. Mahyco conducts large-scale trials on 100 hectares in seven states.

October 2001: Mahyco discovers commercial Bt cotton farming over several thousand hectares in Gujarat. Source of the cotton is traced back to Navbharat Seeds Pvt Ltd.

October 2001: GEAC orders Bt cotton fields to be burnt in Gujarat. No action taken after farmers' protest order.

November 20, 2001: Gene Campaign files a case in the Delhi High Court charging the Government with negligence in allowing large-scale field trials to be conducted without appropriate monitoring, regulation and safety precautions.

January 23, 2002: Dr Manju Sharma, secretary of DBT, declares that the latest round of Bt cotton trials were satisfactory and that it is up to the GEAC and the Ministry of Environment to decide on a date of commercial release.

February 20, 2002: The Indian Council of Agricultural Research (ICAR) submits a positive report to the Ministry of Environment on the field trials of Bt cotton. It is now expected that the Genetic Engineering and Approval Committee (GEAC) of the environment ministry will approve commerical use of Bt cotton within a month.

Bt Cotton: Analysis of Discontent & Suspicion Factors

"We commit to transparency by making published scientific data and data summaries on product safety and benefits publicly available and accessible, and we commit to working within the rigorous science-based regulation as

required by appropriate Government agencies around the world." Monsanto's Corporate Pledge

The plot is uncannily similar. Desperate farmers, pest-ridden crops, a market of endless demand that promises riches to those who can supply, and a magic technology that promises to cure all ills. Less than half a century ago, green revolution was the panacea. Now, its the gene revolution. Gene or green, the proponents of the revolution remain the same multinational chemical companies like Monsanto and Novartis.

Many farming sector observers in India caution that an obvious Government-industry nexus is pushing genetically modified seeds in India without heeding the concerns around the potentially devastating environmental and social impacts of the technology in a largely agrarian economy such as India.

U.S. chemical giant Monsanto's efforts — through its Indian partner Mahyco — to seek approval for commercializing the controversial transgenic Bt Cotton seeds in India has come under intense criticism. Cotton, a pest prone crop, is grown widely in India. Monsanto hopes that its transgenic variety, known as 'bollgard' cotton, can corner this market. Monsanto's cotton carries a gene from a naturally occurring toxic bacterium Bacillus Thurigensis. Splicing this gene within the genetic structure of the cotton makes the cotton plant poisonous to the boll weevil, a notorious cotton pest.

Critics of the Indian Government's handling of the Bt Cotton evaluations say that the Government has condoned serious irregularities and safety violations. They complain that the process lacks transparency and public debate, and that the Government has neither the political will nor the technical and infrastructural ability to monitor or regulate this controversial technology.

After a four-year period of controversy and negotiations shrouded in secrecy, the Government of India is set to approve the commercial planting of Monsanto's Bt Cotton in India.

Monsanto has been operating in India since 1949, and is a market leader in agricultural chemicals. It operates three Indian subsidiaries: Monsanto India, Monsanto Enterprises and Monsanto Chemicals, and early in 1998, Monsanto acquired a 26 per cent stake in the Indian seed company Mahyco. A corporate profile of Monsanto can be downloaded from: http://www.indiaresource.org/issues/agbiotech/2003/www.genewatch.org/GeneSrch/Companies/Monsanto.htm

Fabric of Death

Ironically, it is cotton — touted as the poster child of the failed Green Revolution — that is set to herald the genetic age in Indian agriculture. More than 50 percent of the total pesticides used in India are sprayed on cotton crops. This has exacted a deadly toll both on the farming and on the farmers.

According to New Delhi-based food and trade policy expert Devinder Sharma, more than 10,000 cotton farmers have killed themselves after the introduction of fourth-generation pesticides, called synthetic pyrethroids, less than 20 years ago. Farmers are caught in vicious cycles of debt to meet the high costs of green revolution cotton farming: debts they incur to purchase pesticides, fertilizers, water pumpsets and hybrid seeds. For many, the only way out of the debt trap is to swallow the very pesticides they purchase with their loans.

This year, the cycle of suicides continued as large-scale crop failures floored the already debt-ridden cotton farmers from the Northern state of Punjab to Karnataka in the South. Forced to deliver a quick-fix remedy, the Government of India has latched on to the biotech bandwagon. Estimates place the Indian market for biotechnology at $2.5 billion.

Critics say that...the Government has condoned serious irregularities and safety violations...the process lacks transparency and public debate; and that the Government has neither the political will nor the technical and infrastructural ability to monitor or regulate this controversial technology.

Behind Closed Doors

In April 1998, the first field trials of Bt Cotton were permitted by the Indian Department of Biotechnology's Review Committee on Genetic Manipulation (RCGM).

According to Afsar H. Jafri, program coordinator at Research Foundation for Science, Technology and Ecology, this was a procedural irregularity. The Review Committee, which had authorized the import of Bt seeds in 1995, was empowered only to grant clearances for contained genetic experiments — in laboratories or greenhouses. Permissions for field trials, Jafri says, ought to have been given by the Genetic Engineering Approval Committee.

The Research Foundation, which is run by well-known globalization critic Dr. Vandana Shiva, took the matter to the Indian Supreme Court challenging the allegedly "illegal" field trials.

Even before the case was decided, permission was granted in July 2000 for large-scale field trials this time by the Genetic Engineering Approval Committee (GEAC). Dr. Manju Sharma, secretary of the Department of Biotechnology, said the decision was taken after reviewing the data from the controversial small-scale field trials. According to Dr. Sharma, the data is "totally confidential" but that they allow a "clear inference" from it that Bt is safe.

To date, none of the data generated by Monsanto has been made public. Approval for commercial planting was imminent last year. On June 18, 2001, an open dialogue was called by the GEAC in response to public pressure. The meeting was attended by Greenpeace, scientists, officials and farmers. Questions raised about the scientific aspects of Bt cotton were dismissed and no primary data was used to counter the points of contention raised by Greenpeace.

The Genetic Engineering Approval Committee, however, ordered a one-year extension for field trials of Bt cotton before taking a decision. In a press note , the Committee said the dates of planting Bt cotton were late, in some cases by three months. As a result of the late planting, the pest load was low and the yield data and the net agronomic advantage derived from the study could not reflect true values, Committee members reasoned.

This time around, the trials were to be held under the direct supervision of the Indian Council of Agricultural Research (ICAR). The data from these tests are reportedly with the Government. *[Since this report was written, news accounts indicate that the ICAR has responded positively to the tests, and that the Government will be announcing the commercial use of Bt Cotton anytime now —CWI editor]*

As far as Monsanto is concerned, the ball is in the Government's court. "All trial data has been systematically recorded by Mahyco as required and submitted to the concerned authorities," says Ranjana Smetacek, the director of Government & Public Affairs at Monsanto. "If you are speaking of those Government authorities, I imagine they are governed by a protocol and abide by it." DBT secretary Dr. Sharma refused to talk to CorpWatch India.

The results of the second round of large-scale trials on 100 hectares are yet to see the light of day. But the decision to commercialize Bt Cotton is a fait accompli.

According to Dr. Sharma, the results from the latest round of trials of Bt cotton, commissioned last June, were satisfactory. She told the national daily Indian Express in Bangalore that it is upto the GEAC, and the Ministry of Environment and Forests to decide on a commercialisation date.

A Twist in the Tale

Shortly after public interest groups challenged the Governments haste in fast-tracking Bt Cotton approvals in the Supreme Court, the cotton controversy took an interesting turn. In January 2001, a 10-member U.S. delegation, comprising judges and scientists, met Supreme Court Chief Justice A.S. Anand in New Delhi. According to a report in the national daily *The Hindu*, the object of the meeting with the Chief Justice was to impress upon him and the judicial fraternity the benefits of biotechnology. The delegation was organized by U.S.-based non-profit Einstein Institute for Science, Health and the Courts. Responding to *The Hindu*, Einstein Institute director Dr. Franklin Zweig, who led the delegation, denied that the two-hour meeting was to influence the judiciary. He said it was to educate the judge(s) about the basic principles of public information for use of courts and court systems. The delegation offered to hold workshops for the judges of the Indian Supreme and High courts to educate them about transgenics and safety protocols in biotech research.

The Government, it seems, made up its mind long ago. It has virtually ignored the vigorous anti-GMO campaigns.

In November 1998, farmers from the Southern state of Karnataka burnt down a field where Bt Cotton was being surreptitiously tested by Monsanto-Mahyco. This direct action, which launched the "Operation Cremate Monsanto" campaign, had the full participation of farmer Basanna Hunsole, the owner of the experimental plot.

Hunsole says Monsanto merely told him that the seeds would give him good results and withheld information that the seed was genetically modified. Hunsole adds that the Bt Cotton performed "miserably" in comparison to the traditional varieties planted by him in the nearby plots.

This year, the cycle of suicides continued as large-scale crop failures floored the already debt-ridden cotton farmers...

Vijay Jhawandia of Maharashtra-based farmers organization Shetkari Sanghatana concurs. His visits to field test sites at Wardha in Central India, he says, have left him with the impression that the yields and pest-resistance abilities of Bt Cotton are vastly overrated. "I feel Bt cotton has failed in Maharashtra and there must be an open discussion on this," he said. "Scientists should answer some questions — why is Bt cotton failing here while other varieties are doing well?"

Jhawandia also says that the pests are rapidly developing resistance to the Bt gene. CorpWatch India was unable to verify this claim. However, agricultural columnist Devinder Sharma says that evidence from China and Australia do point to the possibility of increased immunity of pests to Bt.

"It is widely accepted that in the case of Bt Cotton, the third generation of the pest is the most problematic," writes Sharma in an article published by Centre for Alternative Agricultural Media. "It is true that in southern China farmers have been growing this variety. But what is not known is that now they have to spray pesticides to control third and fourth generation of American bollworm insects. In Australia too, farmers have now been advised to go in for more sprays because of a drop in expression levels. With the insect increasingly developing immunity against the Bt toxin in the plant, scientists are now trying to introduce genetically manipulated varieties with two Bt genes."

GM crops in Canada are in danger of causing nuisance weeds, according to a BBC News report citing research by English Nature, a UK Governmental agency that champions wildlife conservation in Britain. The research shows that genetically manipulated "herbicide-resistant oilseed rape is crossbreeding at the edge of fields. The plants are accumulating extra genes and are rapidly becoming resistant to agrochemical sprays, says English Nature." The agency warns that this "could lead to rogue GM crop plants that are harder to control."

Even as the controversy rages on, the Government is causing increasing concern among its critics by exposing its inability, or unwillingness as some say, to monitor and regulate the entry and spread of genetically modified organisms. These concerns take on a serious shade in light of the growing body of evidence pointing to the irreversible effects on genetic pollution on traditional and wild varieties of plants.

Speaking at a conference at the University of Agricultural Sciences in Bangalore in December 1998, Department of Biotechnology secretary Dr. Manju Sharma sought to reassure skeptics: "We have a built-in level of mechanical checks to prevent import of unwanted material."

The first challenge to Sharma's assertion came in June 2001, when a Greenpeace expose pointed to contamination of popular food products such as Pringles potato chips and Isomil baby food with Monsanto's genetically engineered crops. Importing or selling any genetically engineered food products without the prior approval of the Genetic Engineering Approval Committee is illegal under Indian law. Greenpeace confirmed with the Approval Committee that neither Monsanto, nor the manufacturers of Pringles and Isomil, Proctor & Gamble and Abbott Laboratories respectively, had sought or obtained approval from the GEAC. Despite the weight of evidence, the GEAC failed to take any action.

Late last year, the GEAC was caught sleeping yet again. Monsanto-Mahyco tipped off the Government that several thousand hectares of land in Gujarat were already under Bt Cotton production, using seeds distributed by Gujarat-based Navbharat Seeds Pvt Ltd. The Committee ordered that the fields be burnt. However, local farmers, who claimed they did not know that the Navbharat seeds were transgenic, opposed this. In any case, as Rajni Dave of Ahmedabad-based NGO Human Technology Forum points out, the damage may have already been done. More than 100 tons of Bt seeds were collected from cotton ginning factories in Gujarat. No estimate is available of the quantity of Bt seeds that may have already escaped notice.

Gene Campaign, a New Delhi based organisation, has filed a petition in the Delhi High court charging the Government with negligence for allowing large-scale planting of Bt cotton. The petition, which was admitted by the court, demands that GMOs be released only after a rigorous regulatory procedure and an evaluation by an independent regulatory agency (not the government controlled GEAC).

Its proven inabilities notwithstanding, the Government has promised to allow 20 genetically manipulated crops for commercial planting this year. Indian daily Economic Times reports that: "Such crops will range from transgenic cotton and mustard to protein-fortified potatoes and pest-busting tomatoes, Manju Sharma, secretary in the department of biotechnology (BDT), told journalists on the sidelines of the ongoing India Economic Summit of the World Economic Forum."

The evaluation process for these crops will likely begin after the approval is firmed.

Ag-Biotech: An Attempt Towards Making a Forecast

The application of genetic engineering to agriculture continues to hold great commercial potential in the eyes of many. A recent report by BioScience Securities Inc. (BSI), an institutional research and investment banking company, tells of the coming of a commercial ag-biotech revolution, bringing with it the development of novel biopesticides, insect and disease resistant plants, herbicide tolerant crops and premium vegetables (Biotechnology Business News, August 12, 1994). Although some of these products are in late-stage development or already on the market, it is anticipated that a number of agricultural biotechnology firms will move firmly into the commercialization stage during the years 1994-1997.

One of the main thrusts of ag-biotech, according to the BSI report, should be the "formation of new technology-based agricultural business systems, which will be created through the integration of new proprietary biotechnology-derived products with the agribusiness and food processing infrastructure." The formation of this new business system will be driven by a heightened inclination of companies at all points along the commercial continuum (from gene cloners to food distributors) to vertically integrate to capture the value-added benefit created by the use of biotechnology. It is thought that this integration of agricultural biotechnology with downstream agribusiness companies will lead to successful technology-based firms that combine "established dominant market share positions downstream with leading-edge technology positions upstream." Although the story told in the BSI report is positive for the agricultural biotechnology community, it has not yet been heard on Wall Street.

In contrast to the over 289 biotechnology firms that currently exist, which have been formed to focus primarily or secondarily on drug or vaccine development, there are only 74 biotechnology firms that are similarly focused on the development of plant agricultural or food products. Of the 289 therapeutic biotechnology firms, 42 percent are public while only 39 percent remain private. In contrast, of the 74 food and plant ag-biotechnology firms, only 24 percent are public, while over 50 percent are private (Institute for Biotechnology Information, Research Triangle Park, NC, 1994). Of an approximate $465 million raised through IPO's and follow-on offerings by U.S. biotechnology firms between January and July 1994, none of it was raised by food or plant ag-biotechnology firms (Strategic Developments in Biotechnology, September, 1994).

The availability of capital for start-up ag-biotech firms through venture capitalists is no better, according to Ernst & Young's 1995 biotechnology report. Its annual venture capital survey reveals that only two percent of the total venture capital invested in biotechnology went to ag-biotech companies, and this includes animal agriculture. The increased difficulty in raising capital is apparent in the fact that ag-biotech firms have, on average, only enough cash on hand to survive 12 months, down from 38 months the previous year (calculated by comparing cash on-hand to monthly net burn rate)(Lee, K.B. and Burrill, G. S., Biotech

95: Reform, Restructure, Renewal - The Industry Annual Report. Ernst & Young, 1994). As a rule of thumb, firms are considered to be vulnerable if they have less than 24 months cash on-hand.

Given the potential size of the markets catered to by the agricultural biotechnology sector, the success of an early entry product such as Calgene's recombinant tomato may be the flame that thaws the freeze on Wall Street and ignites the beginning of the "agritechnology" era.

Role of Feeding Biotechnology in the World

"Remember, it's about food," noted Cyrus Ndiritu at the conference entitled "Agricultural Biotechnology for a Better World" held April 28-30, 1997 at the Asilomar Center at Pacific Grove, California. Ndiritu, of the Kenya Agricultural Research Institute, outlined his research strategy to approximately 150 participants from around the world.

The meeting brought together researchers, government officials, and business people from around the world to address technical, legal and social issues of biotechnology applied to agriculture. The event was organized by Michigan State University and the United States Agency for International Development (USAID).

The meeting place was apt. Asilomar was the site of the famous 1975 Asilomar Conference on safety and regulation of recombinant DNA technologies discovered in the early 1970's by researchers working 100 miles north in the San Francisco Bay Area. Issues of safety, risk, costs, benefit and regulation raised 22 years earlier are still being wrestled with today.

Humans face the challenge of producing enough food to meet the demands imposed by several biological and agricultural factors:

- Rising population.
- Rising income, and an expectation of a higher quality, more diverse diet.
- Decreasing amount of land available for food production.
- Decreasing the use of synthetic pesticides.
- Preserving biodiversity.

What does it take for a developing nation to identify and solve its own problems using primarily resources from its own people and land? How can nations build their own modern biotechnology community capable of education, training, research, development, and application of biotechnology to increase and improve the food supply?

These are some of the challenges of international "Technology Transfer." Science and knowledge, like news, can spread relatively fast to anywhere, by mail or phone or e-mail. But skill, know-how and technology are slower to transfer. This takes an exchange of ideas and an interchange among people and a commitment of time and money and patience.

Many biosciences teachers in US high schools can understand the challenge of establishing a biotechnology infrastructure. It's one thing to teach about DNA. It's another thing to have year after year the budget, the supplies, the equipment, and the access to expertise and supplies needed to train students to work with DNA and to explore the edge of molecular biology.

Scientists organized this meeting, because scientists run the organizations that sponsor the research featured at the meeting. Yet many students in high school or college might not know that people trained as scientists often end up serving as government officials. Although trained as biological researchers, some develop the skills needed to run a bureaucracy. They need the savvy of the diplomat—a combination of tact, respect and resolve.

Within their own national governments, these scientists/public administrators compete for budgets and personnel and programs. People in departments of agriculture negotiate with their colleagues in the departments of environment over the national stances on regulation, GATT (Global Agreement on Tariffs and Trade) or the Biodiversity Treaty.

Between nations, these scientists/diplomats build partnerships to use biotechnology as a way to improve agriculture, while respecting cultural differences between the cooperating countries.

The great range in biotechnology capacity among countries requires both tact and bluntness by the cooperating countries. Like other people, scientists can't read each others' minds. Kenya's Dr. Ndiritu emphasized that developing countries need to be specific when asking for assistance, and that developed countries need to be straightforward in providing the requested assistance—and let the needs, not the technology, drive the partnership.

Agricultural Biotechnology

Ask your students to examine a globe and compare it to a flat map of the world.

Which is the more accurate model of the world?

What criteria do the students use in comparing the globe to the map?

Globes give a more accurate image than flat maps of physical geography, the scale and location of lands and oceans.

Globes can also help illustrate differences in climates and carrying capacity of the land.

A globe showing national borders is a great tool for learning international politics. Students can better understand the connections between language, law, religion and regulation as affected by history, including colonization, and by current trading patterns, including the size, scope and value of international trade.

Globes are a way to locate where crops originated, and to trace how humans have spread many crops around the world. Potato, tomato, corn, pepper and squash have spread from the New World to the whole world. Oil palm in the past century was introduced to Malaysia from west Africa and South America. Wheat and barley from the eastern Mediterranean grow around the world. These origins and past migrations still influence how food is grown, and how new cultivated varieties are developed.

The center of diversity of a crop is also considered its "ancestral homeland," the place biologists believe the crop originated. Can your students locate the ancestral homes of such key crops listed above?

At this conference, a key issue included the risks and benefits of planting transgenic crops in the center of diversity of that crop. An example would be planting transgenic potatoes in Peru or transgenic corn in Mexico.

The concern is whether transgenic crops significantly threaten landraces of the crop or plants such as teosinte that are wild ancestors of a crop. For example, what is the likely damage if the "transgene" transfers by pollen to native "landraces" of the crop or to a wild ancestor? Would the landraces of the crop become weedy, or weak, or less valuable as a crop to the local farmers? Alternatively, would farmers choose to use the new transgenic varieties, and stop planting seeds of the traditional landraces that have been bred for generations?

The Chips are Coming

Paranoid ranting is a staple on the internet, and the topic of biochips is a particular favorite. Biochips - any microprocessor chips that can be used in biology - mean many things to many people. The web's preachers focus on a simple idea: an implanted chip that identifies each person uniquely, and can be used to track their location.

This wonderful idea - capable of locating lost children, downed soldiers, and wandering Alzheimers patients - sounds alarming even to many mild-mannered civil libertarians. To Living Truth Ministries of Austin, TX, it is no less than "a tremendous bonanza for the coming Antichrist!" Second Advent Ministries of San Bernardino, Calif., warns us that the US social security number will soon be replaced (at the orders of the "New World Order United Nations-controlled world government", no less) by a chip-encoded "18-digit numbering system [which] will consist of THREE separate sets of SIX DIGITS each: 6-6-6! Coincidental? No. Not if you believe Bible prophecy."

Reality is somewhat less alarming. A simple ID chip is already walking around in tens of thousands of individuals, but all of them are pets. Companies such as AVID (Norco, Calif.), Electronic ID, Inc. (Cleburne, Tx.), and Electronic Identification Devices, Ltd. (Santa Barbara, Calif.) sell both the chips and the detectors. The chips are the size of an uncooked grain of rice, small enough to be injected under the skin using a syringe needle. They respond to a signal from the detector, held just a few feet away, by transmitting out an identification number. This number is then compared to database listings of registered pets.

Daniel Man, a plastic surgeon in private practice in Boca Raton, Fla., holds a patent on a more sinister device: a chip that would enable lost humans to be tracked by satellite. He has a large-scale model, but for miniaturization and regulatory approval he will need a substantial amount of cash. "There are some companies that are interested, but nothing is official," he says. "We're not giving up on it yet."

Chips to Monitor Your Every Footstep and Body Fluid

The civil liberties debate over biochips has obscured their more ethically benign and medically useful applications. Medical researchers have been working to integrate chips and people for many years, often plucking devices from well-known electronic appliances. Jeffrey Hausdorff of the Beth Israel Deaconess Medical Center (Boston, Mass.) has used the type of pressure-sensitive resistors found in the buttons of a microwave oven as stride timers. He places one sensor in the heel of a shoe, and one in the toe, and adds a computer to the ankle to calculate the duration of each stride.

"Young, healthy subjects can regulate the duration of each step very accurately," he says. But elderly patients prone to frequent falls have extremely variable stride times, a flag that could indicate the need for more strengthening exercises or a change in medication. Hausdorff is also using the system to determine the success of a treatment for congestive heart failure. By monitoring the numbers of strides that a person takes, he can directly measure the patient's activity level, bypassing the often flawed estimate made by the patient.

Hausdorff's chips are external, but a chip proposed by Sensors for Medicine and Science, Inc. (S4MS; Germantown, Md.) will be injected under the skin. The chip will allow diabetics to easily monitor the level of the sugar glucose in their blood. Diabetics currently use a skin prick and a hand-held blood test, and then medicate themselves with insulin depending on the result. The system is simple and works well, but the need to draw blood means that most diabetics don't test themselves as often as they should. Although they may get away with this in the short term, in later life those who monitored infrequently suffer from blindness, loss of circulation, and other complications.

The solution is more frequent testing, using a less invasive method. Various companies are trying to extract the required body fluid, using either an electric current (Cygnus, Inc., Redwood City, Calif.) or a tiny pore induced by a laser (SpectRx, Norcross, Ga.) or a small needle (Integ, St Paul, Minn.).

"All of these approaches ameliorate the drawing of blood," says George Daniloff, Chief Scientific Officer of S4MS, "but they still require manipulation of skin, and the skin may be perturbed." The S4MS chip, however, will sit underneath the skin, sense the glucose level, and send the result back out by radio-frequency communication.

A light-emitting diode (LED) starts off the detection process. The light that it produces hits a fluorescent chemical: one that absorbs incoming light and re-emits it at a longer wavelength. The longer wavelength of light is then detected, and the result is sent to a control panel outside the body.

Glucose is detected because the sugar reduces the amount of light that the fluorescent chemical re-emits. The more glucose there is, the less light that is detected. S4MS is still developing the perfect fluorescent chemical, but the key design innovation of the S4MS chip has been fully worked out. The idea is simple: the LED is sitting in a sea of the fluorescent molecules. In most detectors the light source is far away from the fluorescent molecules, and the inefficiencies that come with that mean more power and larger devices. The prototype S4MS chip uses a 22μW LED, almost forty times less powerful than the tiny power-on buttons on a computer keyboard. The low power requirements mean that energy can be supplied from the outside, by a process called induction. The fluorescent detection itself does not consume any chemicals or proteins, so the device is self sustaining.

Daniloff has a working model of an oxygen sensor that uses the same layout. With it's current circuitry it is about the size of a large shirt-button, but the final silicon wafer will be less than a millimeter square. "The scale transition is not a challenge," says Daniloff. "The challenge is to illustrate the principle." The oxygen sensor will be useful not only to monitor breathing in intensive care units, but also to check that packages of food, or containers of semiconductors stored under nitrogen gas, remain airtight.

Tom Ferrell of the Oak Ridge National Laboratory in Oak Ridge, Tenn., is also developing an oxygen-sensing chip, but his version sends light pulses out into the body. The light is absorbed to varying extents, depending on how much oxygen is being carried in the blood, and Ferrell's chip detects the light that is left. The rushes of blood pumped by the heart are also detected, so the same chip is a pulse monitor. A number of companies already make large-scale versions of such detectors, and Ferrell says he has "collaborated with almost every company in this area," although he will not name names.

Ferrell estimates that the oxygen chip is two years away, but he has already reduced the dimensions of his temperature-sensing chip to 3mm per side. The transition of certain semiconductors to their conducting state is inherently sensitive to temperature, so designing the sensor was simple enough. With some miniature radio-frequency transmitters, and foam-rubber earplugs to hold the chip in place, the device is complete. Ferrell sees applications for anyone from sick children, to chemotherapy patients who can be plagued by sudden rises in body temperature in response to their anti-cancer drugs.

Brain Surgery with an On-off Switch

Sensing and measuring is one thing, but can we switch the body on and off? Heart pacemakers use the crude approach: large jolts of electricity to synchronize the pumping of the heart. The electric pulses of the Activa implant, made by Medtronic, Inc. (Minneapolis, Minn.), are directed not at the heart but the brain. They turn off brain signals that cause the uncontrolled movements, or tremors, associated with diseases such as Parkinson's.

Drug therapy for Parkinson's disease aims to replace the brain messenger dopamine, a product of the brain cells that are dying. But eventually the drug's effects wear off, and the erratic movements come charging back. "I've had patients tell me that they scare their grandkids because they are flaying around like a fish," says Gary Heit, a neurosurgeon at Stanford University Medical Center. "Imagine your whole body twisted like a charleyhorse."

The next option is to remove a portion of the thalamus, a brain structure that triggers the tremors. The implant, cleared for use in the United States in August 1997, is a new alternative that uses high-frequency electrical pulses to reversibly shut off the thalamus. "It has the same benefits as thalamotomy [the brain operation] without the risks," says Heit. The implantation surgery is far less traumatic than thalamotomy, and if there are any post-operative problems the stimulator can simply be turned off.

Adding Back the Senses

The Activa implant interferes with aberrant brain functioning. The most ambitious bioengineers are trying to add back brain functions, restoring sight and sound where there was darkness and silence.

The success story in this field is the cochlear implant. Most hearing aids are glorified amplifiers, but the cochlear implant is for patients who have lost the hair cells that detect sound waves. For these individuals, no amount of amplification is enough.

The cochlear implant delivers electrical pulses directly to the nerve cells in the cochlea, the spiral-shaped structure that translates sound into nerve pulses. In normal hearing individuals, sound waves set up vibrations in the walls of the cochlea, and hair cells detect these vibrations. High frequency noises (deep notes) vibrate the base of the cochlea, while low frequency notes vibrate nearer the top of the spiral. The implant mimics the job of the hair cells. It splits the frequencies of incoming noises into a number of channels (typically eight), and then stimulates the appropriate part of the cochlea.

The two most successful implants are the Clarion (developed at the University of California at San Francisco (UCSF), Research Triangle Institute, N.C., and Advanced Bionics Corporation of Sylmar, Calif.) and the Nucleus (developed at the University of Melbourne, Australia, and made by Cochlear of Sydney, Australia). Upgrades largely focus on improving the speech processing software, which is operated by a minicomputer worn on the patient's belt. Theoretically, increasing the number of channels (and electrodes) could improve sound perception. But speech is perceived in an area of the cochlea only 14 mm long, and spacing the electrodes too close to each other causes signals to bleed from one channel to another.

The result is a broad-brush version of hearing. While some recipients of the devices report speech-like sounds, many characterize their new world as being populated with legions of quacking ducks or banging garbage cans. But the success is undeniable. "Currently two thirds to three quarters of patients [with more recent models] can understand speech without lip-reading," says Steve Rebscher, a member of the UCSF team. "It's pretty amazing, and certainly better than a lot of people anticipated these devices would do."

With the ear at least partially conquered, the next logical target is the eye. Several groups are working on implantable chips that mimic the action of photoreceptors, the light-sensing cells at the back of the eye. Photoreceptors are lost in both retinitis pigmentosa, a genetic disease, and age-related macular degeneration, the most common cause of lost sight in the developed world.

Joseph Rizzo of the Massachusetts Eye and Ear Infirmary, and John Wyatt of the Massachusetts Institute of Technology have made a twenty-electrode, 1mm-square chip, and implanted it at the back of rabbits' eyes. The original chip, the thickness of a human hair, put too much stress on the eye, so the new version is ten times thinner. The final set-up will include a fancy camera mounted on a pair of glasses. The camera will detect and encode the scene, then send it into the eye as a laser pulse, with the laser also providing the energy to drive the chip. Rizzo has confirmed that his tiny array of light receivers (photodiodes) can generate enough electricity to run the chip. He has also found that the amount of electricity needed to fire a nerve cell into action is ~100-fold lower in the eye than in the ear, so the currents can be smaller, and the electrodes more closely spaced.

For now the power supply comes from a wire inserted directly into the eye and, using this device, Rizzo has detected signals reaching the brain. “We know we can get visual information to the brain,” he says, “but we don’t know what kind of vision that will result in.”

Eugene de Juan of Johns Hopkins Wilmer Eye Institute (Baltimore, Md.) is trying to answer that question by using human subjects. His electrodes, inserted directly into the eye, are large and somewhat crude. But his results have been startling. Completely blind patients have seen well-defined flashes, which change in position and brightness as de Juan changes the position of the electrode or the amount of current. In his most recent experiments, patients have identified simple shapes outlined by multiple electrodes. With as little as an 8 x8 array, de Juan believes he could approximate character recognition, and a 25 x 25 array might give a crude image.

The big money in eye implants is in Germany, where the government has pledged ~$10 million to two projects. One is similar to the US projects, in which chips are implanted on the surface of the retina, the structure at the back of the eye. The other project is putting its implants at the back of the retina, where the photoreceptors are normally found. These ‘subretinal’ chips may block the transport of oxygen and food to the overlying nerve cells, so Eberhart Zrenner of the University of Tübingen, Germany, is developing ‘chain mail’ electrode arrays, with plenty of holes for the delivery of supplies.

The subretinal approach more closely mimics the action of the lost photoreceptors. The chip delivers its message at the source, before intervening layers of retinal nerve cells do some complicated processing. Rizzo thinks that his chip’s hardware will be able to compensate for these computations, but for now he has other things to worry about. "There are all sorts of processes that go on that we're just going to ignore initially," he says. As with the cochlear implant, he says, "the biological tests will show you the best way to stimulate. Ultimately the only thing that matters is what the person sees."

The first big test - showing that a remotely powered device can give true visual signals - is yet to be done. "We haven't got to first base yet," says Rizzo, "but we're close." With a working device many years away, Rizzo is a model of caution. "It's always better to talk down the possibilities than to talk them up, because there are many reasons why the implants might fail," he says. "This is science fiction stuff."

Human cells are almost ridiculously tiny and efficient. Every one of them has an entire genetic instruction set - three billion As, Cs, Gs and Ts - packed into a nucleus one hundredth of a millimeter across. And a few cents worth of carbon, nitrogen and phosphorus will keep millions of cells happy in a dish, reproducing themselves and their DNA data.

Incyte Pharmaceuticals, Inc. (Palo Alto, Calif.) also stores DNA data, but in a way that is somewhat less compact and far more likely to impress a computer geek. Ground-zero for Incyte is a highly air-conditioned room filled with scores of black, towering supercomputers worth up to a million dollars each.

Nine of the ten largest pharmaceutical companies in the world each pay Incyte ~$5 million per year to look at the data on those computers. The money has brought an expanding workforce - from 160 to 675 employees in the last two and a half years - and a frantic effort to keep generating more and more information. "We're maybe a year away from going bankrupt if we don't pursue new technologies at any one point," says Tod Klingler, Director of Research Bioinformatics at Incyte.

The manpower and computer power is needed to tame an ever-growing mountain of DNA sequence information. "We now have 24 organisms completely sequenced we have their entire genetic message written down," says Temple Smith of Boston University. "This information allows us to get information about one organism and use that to understand another."

Paranoid ranting is a staple on the internet, and the topic of biochips is a particular favorite. Biochips - any microprocessor chips that can be used in biology - mean many things to many people. The web's preachers focus on a simple idea: an implanted chip that identifies each person uniquely, and can be used to track their location.

This wonderful idea - capable of locating lost children, downed soldiers, and wandering Alzheimers patients - sounds alarming even to many mild-mannered civil libertarians. To Living Truth Ministries of Austin, TX, it is no less than "a tremendous bonanza for the coming Antichrist!" Second Advent Ministries of San Bernardino, Calif., warns us that the US social security number will soon be replaced (at the orders of the "New World Order United Nations-controlled world government", no less) by a chip-encoded "18-digit numbering system [which] will consist of THREE separate sets of SIX DIGITS each: 6-6-6! Coincidental? No. Not if you believe Bible prophecy."

Reality is somewhat less alarming. A simple ID chip is already walking around in tens of thousands of individuals, but all of them are pets. Companies such as AVID (Norco, Calif.), Electronic ID, Inc. (Cleburne, Tx.), and Electronic Identification Devices, Ltd. (Santa Barbara, Calif.) sell both the chips and the detectors. The chips are the size of an uncooked grain of rice, small enough to be injected under the skin using a syringe needle. They respond to a signal from the detector, held just a few feet away, by transmitting out an identification number. This number is then compared to database listings of registered pets.

Daniel Man, a plastic surgeon in private practice in Boca Raton, Fla., holds a patent on a more sinister device: a chip that would enable lost humans to be tracked by satellite. He has a large-scale model, but for miniaturization and regulatory approval he will need a substantial amount of cash. "There are some companies that are interested, but nothing is official," he says. "We're not giving up on it yet."

The civil liberties debate over biochips has obscured their more ethically benign and medically useful applications. Medical researchers have been working to integrate chips and people for many years, often plucking devices from well-known electronic appliances. Jeffrey Hausdorff of the Beth Israel Deaconess Medical Center (Boston, Mass.) has used the type of pressure-sensitive resistors found in the buttons of a microwave oven as stride timers. He places one sensor in the heel of a shoe, and one in the toe, and adds a computer to the ankle to calculate the duration of each stride. "Young, healthy subjects can regulate the duration of each step very accurately," he says. But elderly patients prone to frequent falls have extremely variable stride times, a flag that could indicate the need for more strengthening exercises or a change in medication. Hausdorff is also using the system to determine the success of a treatment for congestive heart failure. By monitoring the numbers of strides that a person takes, he can directly measure the patient's activity level, bypassing the often flawed estimate made by the patient.

Hausdorff's chips are external, but a chip proposed by Sensors for Medicine and Science, Inc. (S4MS; Germantown, Md.) will be injected under the skin. The chip will allow diabetics to easily monitor the level of the sugar glucose in their blood. Diabetics currently use a skin prick and a hand-held blood test, and then medicate themselves with insulin depending on the result. The system is simple and works well, but the need to draw blood means that most diabetics don't test themselves as often as they should. Although they may get away with this in the short term, in later life those who monitored infrequently suffer from blindness, loss of circulation, and other complications.

The solution is more frequent testing, using a less invasive method. Various companies are trying to extract the required body fluid, using either an electric current (Cygnus, Inc., Redwood City, Calif.) or a tiny pore induced by a laser (SpectRx, Norcross, Ga.) or a small needle (Integ, St Paul, Minn.).

"All of these approaches ameliorate the drawing of blood," says George Daniloff, Chief Scientific Officer of S4MS, "but they still require manipulation of skin, and the skin may be perturbed." The S4MS chip, however, will sit underneath the skin, sense the glucose level, and send the result back out by radio-frequency communication.

A light-emitting diode (LED) starts off the detection process. The light that it produces hits a fluorescent chemical: one that absorbs incoming light and re-emits it at a longer wavelength. The longer wavelength of light is then detected, and the result is sent to a control panel outside the body.

Glucose is detected because the sugar reduces the amount of light that the fluorescent chemical re-emits. The more glucose there is, the less light that is detected. S4MS is still developing the perfect fluorescent chemical, but the key design innovation of the S4MS chip has been fully worked out. The idea is simple: the LED is sitting in a sea of the fluorescent molecules. In most detectors the light source is far away from the fluorescent molecules, and the inefficiencies that come with that mean more power and larger devices. The prototype S4MS chip uses a 22μW LED, almost forty times less powerful than the tiny power-on buttons on a computer keyboard. The low power requirements mean that energy can be supplied from the outside, by a process called induction. The fluorescent detection itself does not consume any chemicals or proteins, so the device is self sustaining.

Daniloff has a working model of an oxygen sensor that uses the same layout. With it's current circuitry it is about the size of a large shirt-button, but the final silicon wafer will be less than a millimeter square. "The scale transition is not a challenge," says Daniloff. "The challenge is to illustrate the principle." The oxygen sensor will be useful not only to monitor breathing in intensive care units, but also to check that packages of food, or containers of semiconductors stored under nitrogen gas, remain airtight.

Tom Ferrell of the Oak Ridge National Laboratory in Oak Ridge, Tenn., is also developing an oxygen-sensing chip, but his version sends light pulses out into the body. The light is absorbed to varying extents, depending on how much oxygen is being carried in the blood, and Ferrell's chip detects the light that is left. The rushes of blood pumped by the heart are also detected, so the same chip is a pulse monitor. A number of companies already make large-scale versions of such detectors, and Ferrell says he has "collaborated with almost every company in this area," although he will not name names.

Ferrell estimates that the oxygen chip is two years away, but he has already reduced the dimensions of his temperature-sensing chip to 3mm per side. The transition of certain semiconductors to their conducting state is inherently sensitive to temperature, so designing the sensor was simple enough. With some miniature radio-frequency transmitters, and foam-rubber earplugs to hold the chip in place, the device is complete. Ferrell sees applications for anyone from sick children, to chemotherapy patients who can be plagued by sudden rises in body temperature in response to their anti-cancer drugs.

Sensing and measuring is one thing, but can we switch the body on and off? Heart pacemakers use the crude approach: large jolts of electricity to synchronize the pumping of the heart. The electric pulses of the Activa implant, made by Medtronic, Inc. (Minneapolis, Minn.), are directed not at the heart but the brain. They turn off brain signals that cause the uncontrolled movements, or tremors, associated with diseases such as Parkinson's. Drug therapy for Parkinson's disease aims to replace the brain messenger dopamine, a product of the brain cells that are dying. But eventually the drug's effects wear off, and the erratic movements come charging back. "I've had patients tell me that they scare their grandkids because they are flaying around like a fish," says Gary Heit, a neurosurgeon at Stanford University Medical Center. "Imagine your whole body twisted like a charleyhorse."

The next option is to remove a portion of the thalamus, a brain structure that triggers the tremors. The implant, cleared for use in the United States in August 1997, is a new alternative that uses high-frequency electrical pulses to reversibly shut off the thalamus. "It has the same benefits as thalamotomy [the brain operation] without the risks," says Heit. The implantation surgery is far less traumatic than thalamotomy, and if there are any post-operative problems the stimulator can simply be turned off.

The Activa implant interferes with aberrant brain functioning. The most ambitious bioengineers are trying to add back brain functions, restoring sight and sound where there was darkness and silence.

The success story in this field is the cochlear implant. Most hearing aids are glorified amplifiers, but the cochlear implant is for patients who have lost the hair cells that detect sound waves. For these individuals, no amount of amplification is enough.

The cochlear implant delivers electrical pulses directly to the nerve cells in the cochlea, the spiral-shaped structure that translates sound into nerve pulses. In normal hearing individuals, sound waves set up vibrations in the walls of the cochlea, and hair cells detect these vibrations. High frequency noises (deep notes) vibrate the base of the cochlea, while low frequency notes vibrate nearer the top of the spiral. The implant mimics the job of the hair cells. It splits the frequencies of incoming noises into a number of channels (typically eight), and then stimulates the appropriate part of the cochlea.

The two most successful implants are the Clarion (developed at the University of California at San Francisco (UCSF), Research Triangle Institute, N.C., and Advanced Bionics Corporation of Sylmar, Calif.) and the Nucleus (developed at the University of Melbourne, Australia, and made by Cochlear of Sydney, Australia). Upgrades largely focus on improving the speech processing software, which is operated by a minicomputer worn on the patient's belt. Theoretically, increasing the number of channels (and electrodes) could improve sound perception. But speech is perceived in an area of the cochlea only 14 mm long, and spacing the electrodes too close to each other causes signals to bleed from one channel to another.

The result is a broad-brush version of hearing. While some recipients of the devices report speech-like sounds, many characterize their new world as being populated with legions of quacking ducks or banging garbage cans. But the success is undeniable. "Currently two thirds to three quarters of patients [with more recent models] can understand speech without lip-reading," says Steve Rebscher, a member of the UCSF team. "It's pretty amazing, and certainly better than a lot of people anticipated these devices would do."

With the ear at least partially conquered, the next logical target is the eye. Several groups are working on implantable chips that mimic the action of photoreceptors, the light-sensing cells at the back of the eye. Photoreceptors are lost in both retinitis pigmentosa, a genetic disease, and age-related macular degeneration, the most common cause of lost sight in the developed world.

Joseph Rizzo of the Massachusetts Eye and Ear Infirmary, and John Wyatt of the Massachusetts Institute of Technology have made a twenty-electrode, 1mm-square chip, and implanted it at the back of rabbits' eyes. The original chip, the thickness of a human hair, put too much stress on the eye, so the new version is ten times thinner. The final set-up will include a fancy camera mounted on a pair of glasses. The camera will detect and encode the scene, then send it into the eye as a laser pulse, with the laser also providing the energy to drive the chip. Rizzo has confirmed that his tiny array of light receivers (photodiodes) can generate enough electricity to run the chip. He has also found that the amount of electricity needed to fire a nerve cell into action is ~100-fold lower in the eye than in the ear, so the currents can be smaller, and the electrodes more closely spaced.

For now the power supply comes from a wire inserted

directly into the eye and, using this device, Rizzo has detected signals reaching the brain. "We know we can get visual information to the brain," he says, "but we don't know what kind of vision that will result in."

Eugene de Juan of Johns Hopkins Wilmer Eye Institute (Baltimore, Md.) is trying to answer that question by using human subjects. His electrodes, inserted directly into the eye, are large and somewhat crude. But his results have been startling. Completely blind patients have seen well-defined flashes, which change in position and brightness as de Juan changes the position of the electrode or the amount of current. In his most recent experiments, patients have identified simple shapes outlined by multiple electrodes. With as little as an 8 x8 array, de Juan believes he could approximate character recognition, and a 25 x 25 array might give a crude image.

The big money in eye implants is in Germany, where the government has pledged ~$10 million to two projects. One is similar to the US projects, in which chips are implanted on the surface of the retina, the structure at the back of the eye. The other project is putting its implants at the back of the retina, where the photoreceptors are normally found. These 'subretinal' chips may block the transport of oxygen and food to the overlying nerve cells, so Eberhart Zrenner of the University of Tübingen, Germany, is developing 'chain mail' electrode arrays, with plenty of holes for the delivery of supplies.

The subretinal approach more closely mimics the action of the lost photoreceptors. The chip delivers its message at the source, before intervening layers of retinal nerve cells do some complicated processing. Rizzo thinks that his chip's hardware will be able to compensate for these computations, but for now he has other things to worry about. "There are all sorts of processes that go on that we're just going to ignore initially," he says. As with the cochlear implant, he says, "the biological tests will show you the best way to stimulate. Ultimately the only thing that matters is what the person sees."

The first big test—showing that a remotely powered device can give true visual signals - is yet to be done. "We haven't got to first base yet," says Rizzo, "but we're close." With a working device many years away, Rizzo is a model of caution. "It's always better to talk down the possibilities than to talk them up, because there are many reasons why the implants might fail," he says. "This is science fiction stuff."

REFERENCES

Atkinson, M. A. and Leiter, E. H., "The NOD mouse model of type 1 diabetes: As good as it gets?" Nature Medicine 5: 601-604 (1999).

Baker, K. S., *et al.*, "New Malignancies After Blood or Marrow Stem-Cell Transplantation in Children and Adults: Incidence and Risk Factors," Journal of Clinical Investigation 21: 1352-1358 (2003).

Chen, Y., *et al.*, "Embryonic stem cells generated by nuclear transfer of human somatic nuclei into rabbit oocytes," Cell Research 13: 251-264 (2003).

Childress, J., "Federal Policy Toward Embryonic Stem cell research," *American Journal of Bioethics* 2(1): 34 (2002).

Choi K, Kennedy M, Kazarov A, Papadimitriou, Keller G. A common precursor for hematopoietic and endothelial cells. Development 1998;125:725-732.

Faden, *et al.*, "Public Stem Cell Banks: Considerations of Justice in Stem Cell Research and Therapy" *Hastings Center Report* 33:6 (2003).

Faraci, M., *et al.*, "Severe neurologic complications after hematopoietic stem cell transplantation in children," Neurology 59: 1895-1904 (2002).

Jaffredo T, Gautier R, Eichmann A, Dieterlen-Lievre F. Intraaortic hemopoietic cells are derived from endothelial cells during ontogeny. Development 1998;125:4575-4583.

James, Clive, "Global Review of Commercialized Transgenic Crops: 1998," *ISAAA*, No. 8, 1998.

Kahn, J., "Missing the mark on stem cells," *Bioethics Examiner* 5:3, Fall 2001, p. 1.

Kalka, C. et al. Transplantation of ex vivo expanded endothelial progenitor cells for therapeutic neovascularization. Proc. Natl. Acad. Sci. USA 2000; 97:3422-3427.

Kass, L., "The Meaning of Life - In the Laboratory," *The Public Interest*, Winter 2002. Also see the Council's July 2002 report *Human Cloning and Human Dignity: An Ethical Inquiry*, Chapter 6.

Kawano H, Do YS, Kawano Y, Starnes V, Barr M, Law RE, Hsueh WA. Angiotensin II has multiple profibrotic effects in human cardiac fibroblasts. Circulation 2000;101:1130-1137.

Kennedy M, Firpo M, Choi K, Wall C, Robertson S, Kabrun N, Keller G. A common precursor for primitive erythropoiesis and definitive haematopoiesis.

Kennedy, D., "Stem Cells: Still Here, Still Waiting," *Science* 300: 865 (2003).

Kerr, D. A., et al., "Human Embryonic Germ Cell Derivatives Facilitate Motor Recovery of Rats with Diffuse Motor Neuron Injury," The Journal of Neuroscience 23: 5131-5140 (2003).

Kerr, D. A., *et al.*, "Human Embryonic Germ Cell Derivatives Facilitate Motor Recovery of Rats with Diffuse Motor Neuron Injury," The Journal of Neuroscience 23: 5131-5140 (2003).

Labastie M-C, Cortes F, Romeo P-H, Dulac C, Peault B. Molecular identity of hematopoietic precursor cells emerging in the human embryo. Blood 1998;92:3624-3635.

Latham, S., "Ethics and Politics," *American Journal of Bioethics* 2(1): 46 (2002).

Lechner, A. and Habener, J.F., "Stem/progenitor cells derived from adult tissues: potential for the treatment of diabetes mellitus," American Journal of Physiology—Endocrinology and Metabolism 284: E259-266 (2003).

Orlic D, Kajstura J, Chimenti S, et al. Bone marrow cells regenerate infarcted myocardium. Nature 2001, 410:701-705.

Orr, R., "The Moral Status of the Embryonal Stem Cell: Inherent or Imputed?" *American Journal of Bioethics* 2(1): 57-59 (2002).

Pagano, S.F., *et al.*, "Isolation and characterization of neural stem cells from the adult human olfactory bulb," Stem Cells 18: 295-300 (2000).

Pevny, L., and Rao, M.S., "The stem-cell menagerie," Trends in Neurosciences 26: 351-359 (2003).

Strong, C., "Those Divisive Stem Cells: Dealing with Our Most Contentious Issues," *American Journal of Bioethics* 2(1): 39-40 (2002).

Zhao, M., et al., "Amelioration of streptozotocin-induced diabetes in mice using human islet cells derived from long-term culture in vitro," Transplantation 73: 1454- 1460 (2002).

Zoloth, Laurie, "Jordon's Banks: A View from the First Years of Human Embryonic Stem Cell Research." In The Human Embryonic Stem Cell Debate. Ed. by Suzanne Holland, Karen Lebacqz, and Laurie Zoloth. Cambridge, MA: The MIT Press, 2001. 238.

Zulewski, H., *et al.*, "Multipotential Nestin-Positive Stem Cells Isolated From Adult Pancreatic Islets Differentiate Ex Vivo Into Pancreatic Endocrine, Exocrine and Hepatic Phenotypes," Diabetes 50: 521-533 (2001).

3

Biotechnology in Forestry and Animal Agriculture

Biotechnology in Forestry, Including Genetic Modification

At its 13th Session, in November 2003, the FAO Panel of Experts on Forest Gene Resources reviewed the relevance and discussed the feasibility of a global review of biotechnology in forestry, as part of an FAO effort to assess the global status and trends of forest tree genetic diversity. This paper is the first attempt to provide statistical information on the extent and patterns of biotechnology research and applications in forest trees, worldwide. Other studies are anticipated, in particular analyses of economic and other forces driving biotechnology development in forestry.

The Forest Genetic Resources Working Papers report on issues and activities related to the conservation, sustainable use and management of forest tree genetic diversity. The purpose of these papers is to provide early information on on-going developments in the field, and to stimulate discussion.

The designations employed and the presentation of material in this publication do not imply the expression of any opinion whatsoever on the part of the Food and Agriculture Organization of the United Nations concerning the legal status of any country, territory, city or area or of its authorities, or concerning the delimitation of its frontiers or boundaries. As a Working Paper, this document does not necessarily reflect the views of FAO.

Applications of Biotechnology

Modern biotechnology techniques are currently being used in many areas such as food, agriculture, health care, forestry, the environment, minerals, oil and gas recovery and industrial processes to develop new products and processes, and to modify existing ones. Summarized below are some of the ways that modern biotechnology techniques have been applied in these areas.

Food and Agriculture

One of the most extensive applications of biotechnology has been in agriculture. Biotechnology techniques such as recombinant DNA techniques and mutagenesis have been used to develop plants with novel traits. These traits include herbicide tolerance and pest, insect and virus resistance. Biotechnology techniques have been used to produce biopesticides which are toxic to targeted plant pests. Also, experiments into genetic modification of aquatic organisms such as salmon, for such novel traits as enhanced growth, have been carried out using recombinant DNA techniques.

Health Care

To date, applications of biotechnology in health care have focused on fighting disease using the human body's own 'weapons'. Biotechnology medicines and therapies synthesize proteins, enzymes, antibodies and other substances which occur naturally in the human body, to fight infections and diseases. However, biotechnology also uses other living organisms, that is, plant and animal cells, viruses and yeasts to help produce human medicines. There are four main areas in health care in which biotechnology is currently being used: medicines, vaccines, diagnostics and gene therapy.

Several biotechnology medicines have been developed and approved to treat such diseases as anaemia, growth deficiency in children, hemophilia, diabetes and others. Biotechnology companies have also developed therapeutic products for infectious agents such as HIV, hepatitis B and influenza. Biotechnology is responsible for new vaccines such as the hepatitis B vaccine which was approved in the U.S. to fight the hepatitis B virus. Biotechnology diagnostics have been used to detect a wide variety of diseases and genetic conditions. For example, gene therapy is used diagnostically to screen donated blood to protect the blood supply from HIV and hepatitis. Biotechnology diagnostics are also used in home pregnancy tests. In a home pregnancy test kit, a protein called a monoclonal antibody (MAb), binds to Human Chorionic Gonadotrophin (HCG), causing a colour change. HCG is present in a woman's urine only during pregnancy. In gene therapy, genes are used as 'drugs' to treat hereditary genetic disorders. Faulty or missing genes can be replaced to prevent the occurrence of a genetic disease. Current gene therapy is primarily experiment-based, with a few human clinical trials, such as for the treatment of cystic fibrosis, in the early stages.

Environment

Biotechnology applications in the environment focus on using living organisms primarily to treat waste and prevent pollution. Examples of these applications include bio-filtration and bio-remediation. Bio-filtration refers to the use of microorganisms to remove complex pollutants from the air emissions and waste water discharges of many manufacturing processes. Bio-remediation refers to a number of processes that use living organisms to degrade toxic waste into harmless by-products such as water, carbon dioxide and other materials. Examples of bio- remediation processes include:

- *Bio-stimulation:* A technique which involves introducing nutrients to stimulate the growth of waste-eating microorganisms already present in the environment at a waste site.
- *Bio-augmentation:* A technique whereby microorganisms, not normally present in the ecosystem of a contaminated site, are added to the site to clean up pollutants. The microorganisms added may be natural or genetically modified. The organisms die when their food is used up.
- *Phytoremediation:* A technique which uses certain plants and fungi, planted at waste sites, to naturally reclaim the sites without using chemical treatment, incineration or land filling. These plants and fungi are used because they flourish by accumulating metals and other waste materials present in the soil.

Forestry

To date, only a few biotechnology-derived products and processes used in forestry have been commercialized. Most biotechnology applications in forestry are still in the research and development stage. However, some commercial applications include the development of bacterial bio-pesticides such as *Bacillus thuringiensis* (Bt), as an alternative to chemical pesticides. Tissue culture technology is also being used to produce genetically modified seedlings for forest regeneration. For example, BCRI Inc., a biotechnology company in British Columbia, is using tissue culture to produce weevil-resistant Sitka spruce. Companies are also developing bio-fertilizers such as plant growth-promoting rhizobia bacterial fertilizers. In addition, genetic engineering has been used to develop traits of agronomic interest such as pest and disease resistance, within trees. The pulp and paper industry has used enzymes and microorganisms in their bio-bleaching and de-inking processes.

Industrial Processes

Biotechnology has been applied in a variety of industrial processes in different ways, particularly in the use of biocatalysts in manufacturing processes. Biocatalysts are substances that initiate or modify the rate of a biological process and are generally consumed in the process. Some examples of industrial processes where biotechnology has been applied include chemicals, starch/grain processing, cleaning and textile industrial processes. In the chemicals industry, biotechnology has been used to produce commodity and specialty chemicals. It is also used in the starch and grain processing industries through the use of enzymes to turn starch into glucose and fructose. Corn and other grains can be converted to sweeteners such as high-fructose corn syrup and maltose syrup, using enzymes. Biotechnology applications have been used to produce ethanol from grain. Furthermore, biotechnology is used in the textile industry for the finishing of fabrics and garments. In the pulp and paper industry, bio-pulping is used to manufacture some products.

Metals and Minerals Recovery

Biotechnology has been used in the mining industry in such processes as bio-leaching and metal bio-remediation and recovery. Bio-leaching is the use of bacteria to extract valuable metals. Metal bio-remediation and recovery refers to the use of biologically created enzymes in the degreasing process to recover metals of value.

Energy

To date, biotechnology applications in the energy industry include the production of cleaner coal and petroleum by removing sulfur and thus reducing the environmental contaminants released during combustion. Biotechnology has also been used widely to produce bio fuels such as bio-ethanol and bio-diesel. Bio fuels are alcohols, ethers, esters and other organic chemicals made from biomass such as herbaceous and woody plants, agricultural and forestry residues and industrial waste. These fuels can be used for electricity and as fuels for transportation.

Human Applications

When treating problems that arise from genetic disorder, one solution is gene therapy. A genetic disorder is a situation where some genes are missing or faulty. When this happens, genes may be expressed in unfavorable ways or not at all, and this generally leads to further complications. The idea of gene therapy is that a non-pathogenic virus or other delivery system can be used to insert a piece of DNA—a good copy of the gene—into cells of the living individual. The modified cells would divide as normal and each division would produce cells that express the desired trait. The result would be that he/she would then have the ability to express the trait that was previously absent at least partially. This form of genetic engineering could help alleviate many problems, such as diabetes, cystic fibrosis, or other genetic diseases. The potential of genetic engineering to cure medical conditions opens the question of exactly what such a condition is. Some advocates see aging and death as medical conditions and engineering problems to be solved. They see human genetic engineering as a key tool in this *(see life extension)*. The difference between cure and enhancement from their perspective is merely one of degree. Theoretically genetic engineering could be used to drastically change people's genomes which could enable people to regrow limbs, the spine, the brain. It could also be used to make people stronger, faster, smarter, or to increase the capacity of the lungs, among other things. If a gene exists in nature, it could be brought over to a human cell. Others feel that there is an important distinction between using genetic technologies to treat those who are suffering and to make those who are already healthy superior to others. There is widespread agreement that germline engineering should not currently be allowed for either therapeutic and enhancement applications, as evidenced by a recent report by the American Association for the Advancement of Science. Though theory and speculation suggest that genetic engineering could be used to make people stronger, faster, smarter, or to increase lung capacity, the AAAS report finds that there is little evidence that this currently can be done safely or without unethical human experiments.

"Transgenic organism" is now the preferred term for genetically modified organisms with extra-genome (foreign genetic) information, as opposed to "genetically engineered" or "genetically modified" organisms (which may refer to changes made within the genome such as amplification or deletion of genes). The first Genetically Engineered drug was human insulin approved by the USA's FDA in 1982. Another early application of GE was to create human growth hormone as replacement for a drug that was previously extracted from human cadavers. In 1986 the FDA approved the first genetically engineered vaccine for humans, for hepatitis B. Since these early uses of the technology in medicine the use of the GE has expanded to supply many drugs and vaccines. One of the best known applications of genetic engineering is that of the creation of genetically modified organisms (GMOs). There are potentially momentous biotechnological applications of GM, for example oral vaccines produced naturally in fruit, at very low cost. A radical ambition of some groups is human enhancement via genetics, eventually by molecular engineering. DNA sequencing is a technique which is used to identify each base in DNA. Although the costs of DNA sequencing has dropped dramatically, the NIH estimates it costs at least $10 million to sequence 3 billion base pairs—the size of the whole human genome. Although there has been a tremendous revolution in the biological sciences in the past twenty years, there is still a great deal that remains to be discovered. The completion of the sequencing of the human genome, as well as the genomes of most agriculturally and scientifically important plants and animals, has increased the possibilities of genetic research immeasurably. Expedient and inexpensive access to comprehensive genetic data has become a reality, with billions of sequenced nucleotides already online and annotated. Now that the rapid sequencing of arbitrarily large genomes has become a simple, if not trivial affair, a much greater challenge will be elucidating function of the extraordinarily complex web of interacting proteins, dubbed the proteome, that constitutes and powers all living things. Genetic engineering has become the gold standard in protein research, and major research progress has been made using a wide variety of techniques, including:

- Loss of function, such as in a knockout experiment, in which an organism is engineered to lack the activity of one or more genes. This allows the experimenter to analyze the defects caused by this mutation, and can be considerably useful in unearthing the function of a gene. It is used especially frequently in developmental biology. A knockout experiment involves the creation and manipulation of a DNA construct in vitro, which, in a simple knockout, consists of a copy of the desired gene which has been slightly altered such as to cripple its function. The construct is then taken up by embryonic stem cells, where the engineered copy of the gene replaces the organism's own gene. These stem cells are injected into blastocysts, which are implanted into surrogate mothers. Another method, useful in organisms such as Drosophila (fruit fly), is to induce mutations in a large population and then screen the progeny for the desired mutation. A similar process can be used in both plants and prokaryotes.

- Gain of function experiments, the logical counterpart of knockouts. These are sometimes performed in conjunction with knockout experiments to more finely establish the function of the desired gene. The process is much the same as that in knockout engineering, except that the construct is designed to increase the function of the gene, usually by providing extra copies of the gene or attracting more frequent transcription.
- 'Tracking' experiments, which seek to gain information about the localization and interaction of the desired protein. One way to do this is to replace the wild-type gene with a 'fusion' gene, which is a juxtaposition of the wild-type gene with a reporting element such as Green Fluorescent Protein (GFP) that will allow easy visualization of the products of the genetic modification. While this is a useful technique, the manipulation can destroy the function of the gene, creating secondary effects and possibly calling into question the results of the experiment. More sophisticated techniques are now in development that can track protein products without mitigating their function, such as the addition of small sequences which will serve as binding motifs to monoclonal antibodies.

Related Technologies and Biotech Industries

This looks at how biotechnology contributes to new and enhanced industrial processes. It examines how biotechnology enables the use of renewable resources in industry, thereby increasing efficiency, decreasing pollution, and reducing energy usage and waste production during various types of industrial processing.

- Biochemicals
- Bioplastics and Biopolymers
- Biotechnology in the Textiles Industry
- Enzymes in Industrial Applications
- Microbial Enhanced Oil Recovery
- Starch Bioprocessing

(a) Biochemicals

Biochemicals are chemicals produced from organic sources known as biomass. They are sometimes called bio-based chemicals, green chemicals, and plant-based chemicals. Biochemicals are chemicals produced from organic sources known as biomass. Biomass is any living matter that can be naturally and regularly replenished, including agricultural food and feed crop residues, aquatic plants, animal wastes and other waste materials. Biochemicals are sometimes called bio-based chemicals, green chemicals, and plant-based chemicals. Currently, the most popular source for biochemicals is corn. Chemicals like ethyl lactate, succinic acid, and polylactic acid are made from corn components, such as corn starch, and are used to make a variety of products, such as plastics, solvents, clothing fibres, paints, and food additives.

Petroleum is the oil found underground in certain parts of the world and is the most used energy source in the world. It is used to make gasoline, chemicals, plastics and drugs. Because petroleum is limited in quantity, and cannot be made once the supply has been depleted from the earth, it is said to be non-renewable. Petroleum products are also not environmentally friendly – causing air pollution when burned, and filling up landfills when disposed of. Biochemical products made from biomass help solve many of the problems associated with petrochemical products. Environmental benefits include less pollution from crude oil extraction and processing, as well as better waste management since the materials are biodegradable. And because biomass can be replenished regularly by nature, biochemical products are said to be renewable.

- *Commodity chemicals — These include solvents, fuel additives, lubricants, surfactants, adhesives, and inks.*
- *Fine chemicals — These include enzymes, nutraceuticals, and pharmaceuticals.*
- *Chemical intermediates — These include sugars, organic acids, and other monomers.*

Biochemicals are made from the processing of plant sugars. Currently, corn is being used extensively as a starting material for biochemicals. Making biochemicals involves two basic steps:

- First, organic matter, such as corn, is milled and the sugar is extracted from the raw material. The sugar is then fermented in a controlled environment, much like the fermentation of beer and wine. Some fermentation processes use genetically modified organisms that convert corn-derived sugar to a specific chemical.
- Next, a purification process is undertaken where the fermented liquid is purified for the target

biochemical through methods such as distillation and electrodialysis. The biochemicals are then made into various products, such as plastics and solvents.

The role of biotechnology in the creation of biochemicals is crucial. Through fermentation technology, biotechnology enables the processing of common crops and crop wastes, such as corn kernels and corn stalks, to make products such as plastics, textiles, and ethanol. Biotechnology also allows the creation of genetically modified bacteria that can convert plant sugars into an ingredient used in clothing, packaging, and plastics. New enzymes are also being developed for breaking down plant sugars. Biotechnology also helps scientists better understand the biology of bacteria and enzymes that can be used in bioprocessing, which may enable the development of cheaper methods of biochemical production. Currently, biochemical products are often expensive to produce and are not cost-effective enough to compete with traditional products.

esearch in biochemicals continues to grow rapidly. Many large multinational companies are investing time and money into producing chemicals that can be economically viable and environmentally friendly. For example, many chemical companies are teaming up with biotechnology companies to develop products based on biochemicals. Cargill Dow currently makes polylactic acid from corn kernels but has plans to switch to cheaper feedstocks, such as corn stalks, wheat straw, rice hulls, and sawdust, so that biochemicals can be more competitive in the marketplace. Current research is also focussed on developing genetically modified microorganisms for use in specific chemical productions. Pollution due to the use of petroleum-based products may be eliminated by the use of biochemicals. Whereas petroleum-based products do not biodegrade and accumulate in waste landfills, biochemicals degrade completely and do not contribute to waste problems. As well, biochemicals are made from biomass; therefore, their production and use are sustainable.

(b) Biopolymers and Bioplastics

Many everyday products used by Canadians are made of plastic. Pens, computers, cars, food packaging, and clothing are all examples of products which contain plastic. Plastic is a material made up of one or more polymers. The main source of the chemicals needed to manufacture plastics are fossil fuels. These petrochemical plastics are very durable, but take a long time to biodegrade when disposed. Rising concern about the cost of fossil fuels, and their impact on the environment has resulted in a search for alternatives to petrochemical plastics, namely biopolymers and bioplastics.

Biopolymers and bioplastics go by many different names. They are often referred to as bio-based plastics and polymers, or as biodegradable plastics or polymers. They are defined below:

- Biopolymers are polymers which are present in, or created by, living organisms. These include polymers from renewable resources that can be polymerized to create bioplastics.
- Bioplastics are plastics manufactured using biopolymers, and are biodegradable.

Biopolymers and bioplastics are not new products. Henry Ford developed a method of manufacturing plastic car parts from soybeans in the mid-1900s. However, World War II side-tracked the production of bioplastic cars. Today, bioplastics are gaining popularity once again as new manufacturing techniques developed through biotechnology are being applied to their production.

There are two main types of biopolymers: those that come from living organisms; and, those which need to be polymerized but come from renewable resources. Both types are used in the production of bioplastics. These biopolymers are present in, or created by, living organisms. These include carbohydrates and proteins. These can be used in the production of plastic for commercial purposes.

(c) Biotechnology in the Textiles Industry

Biotechnology has impacted the textiles industry through the development of more efficient and more environmentally friendly manufacturing processes, as well as through the design of improved textile materials. Some of biotechnology's key roles have involved the implementation, production, and modification of enzymes for the improvement of textile manufacturing processes. Biotechnology has also facilitated the production of novel and biodegradeable fibres from biomass feedstocks.

Enzyme Biotechnology in Textiles

Through biotechnology, enzymes are used to treat and modify fibres during textile manufacturing, processing, and in caring for the product afterwards. Some applications include:

- De-sizing of cotton – Untreated cotton threads can break easily when being woven into fabrics. To prevent this breakage, they are coated with a jelly-like substance through a process called sizing. However, after the threads have been woven into fabrics, the agents needed to further finish the material cannot adhere to the jelly-coated fabrics. Thus, the protective sizing agents must be removed by a process called de-sizing. Amylase enzymes are widely used in de-sizing, as they do not weaken or affect cotton fibres, nor do they harm the environment.

- Retting of flax – Flax plants are an important source of textile fibres. Useful flax fibres are separated from the plant's tough stems through a process called retting. Traditional retting methods consume large quantities of water and energy. Bacteria, which may be bred or genetically engineered to contain necessary enzymes, can be used to make this a more energy efficient process.
- Breakdown of hydrogen peroxide – When cotton is bleached, a chemical called hydrogen peroxide, which can react with other dyes, remains on the fabric. Catalase enzymes specifically break down hydrogen peroxide and may be used to remove this reactive chemical before further dyeing.
- Biostoning and Biopolishing – Instead of using abrasive tools like pumice stones to create a stonewashed effect or to remove surface fuzz, cellulase enzymes may be used to effectively stonewash and polish fabrics without abrasively damaging the fibres.
- Detergents – Enzymes allows detergents to effectively clean clothes and remove stains. They can remove certain stains, such as those made by grass and sweat, more effectively than enzyme-free detergents. Without enzymes, a lot of energy would be required to create the high temperatures and vigorous shaking needed to clean clothes effectively. Enzymes used in laundry detergents must be inexpensive, stable, and safe to use. Currently, only protease and amylase enzymes are incorporated into detergents. Lipase enzymes break down too easily in washing machines to be very useful in detergents. However, their

stability is being studied and further developed through methods such as genetic screening and modification.

Synthetic fibres made from renewable sources of biomass are environmentally sustainable, and are becoming increasingly economically sustainable. Biodegradable synthetic polymers include novel fibres such as polyglycolic acid and polylactic acid, which are made from natural starting materials.

Not all novel fibres are synthetic; they may also be naturally derived. Some natural biological fibres come from basic materials found in nature, including:

- Chitin – a type of sugar polymer found in crustaceans
- Collagen – a type of protein found in animal connective tissue
- Alginate – a type of sugar polymer found in certain bacteria

A prime example of a synthetic biomass fibre is Polylactic Acid (PLA), which is made by fermenting cornstarch or glucose into lactic acid, and then chemically transforming it into a polymer fibre. With properties similar to other synthetic fibres, PLA based materials are durable with a silky feel, and may be blended with wool or cotton.

PLA has potential applications in several areas, including the following:

- Textiles – clothing, fashions, upholstery
- Agriculture – plant mats, tree nets, soil erosion control products
- Sanitation – household wipes, diaper products
- Medicine – disposable garments, medical textiles

PLA minimizes environmental waste, as it may be fully biodegraded by microorganisms under appropriate conditions into carbon dioxide and water. Unlike the non-renewable petroleum resources used to make traditional synthetic fibres, the supply of renewable corn biomass needed to make PLA is expected to surpass demand in the anticipated future.

Biodegradable synthetic fibres and natural biological fibres may be used to make textiles for medical applications. Such textiles may be used in first aid, clinical, and hygienic practices. Some examples are described below:

Polymer Use(s)

Polylactic Acid and Polyglycolic Acid Used in sutures, absorbable wound closure products, orthopaedic repair absorbable pins, and fixation devices, as well as in tissue engineering structures

Collagen Uses in cell engineering structures, such as in artificial skin, or even as surgeon´s thread

Synthetic biomass fibres may also be used in drug delivery systems, which are designed to release drugs at a specified rate for a specified time.

(d) Enzymes in Industrial Applications

Enzymes are proteins that speed up biochemical reactions without being consumed or changed by the reaction. They are found throughout nature; in our bodies, in the environment, and in all living things. Without enzymes, life would not be possible. Enzymes speed up chemical and biochemical reactions, and this process is called catalysis. The substances upon which enzymes work when performing catalysis are known as substrates, and each enzyme will only fit and act upon a specific set of substrates.There are so many enzymes that it would be impossible to name them all. In fact, scientists have yet to discover many enzymes, or fully understand their structure and properties. On the other hand, many other enzymes have been successfully studied and applied to industrial and commercial uses. A few basic enzyme types are briefly described as follows:

Type	*Enzyme Function*
Cellulase	Breaks down cellulose, a fibre found in the cell walls of all plants and trees. Cellulose is the basic raw material used to make products such as paper, cotton, and other textiles
Hemicellulase	Breaks down hemicellulose, another plant sugar that is not as complex as cellulose and is easier to break down
Xylanase	Breaks down xylan, a gummy sugar present in the cell walls of plants and trees. This enzyme type is used primarily in the wood and pulp industry
Amylase	Breaks down starches and other carbohydrates into basic sugars
Protease	Breaks down proteins
Lipases	Breaks down fats

In the absence of enzymes, chemical reactions occur only when molecules collide while in proper alignment with each other. Because molecules are bumping into each other randomly, chemical reactions are essentially due to chance events. This sometimes results in reactions that occur very slowly, or reactions that do not occur at all.Enzymes act like tiny molecular machines to ensure that molecules come into contact with each other and react. Like a key fitting into a lock, chemical molecules fit into pocket-like structures located on an enzyme. These pockets hold the molecules in a position that will allow them to react with each other, ensuring that they are close enough together and aligned properly for a reaction to occur.

In this way, enzymes speed up reactions. The enzymes are not changed themselves by the reaction. When the reaction is complete, enzymes release the product(s) and are ready to bring together more molecules and catalyse more reactions. Biotechnology and EnzymesBiotechnology can provide an unlimited and pure source of enzymes as an alternative to the harsh chemicals traditionally used in industry for accelerating chemical reactions. Enzymes are found in naturally occurring microorganisms, such as bacteria, fungi, and yeast, all of which may or may not be genetically modified.

Large quantities of enzymes are often needed for industrial use, so these microorganisms are multiplied through a process called fermentation. When enzymes or the microorganisms used to create them are no longer needed, they are destroyed through exposure to heat or safe organic/inorganic materials.

Enzyme Biotechnology in the Pulp and Paper Industry

Enzymes have been used in the pulp and paper industry to soften wood fibres, improve drainage, and present alternatives to chemical bleaching.

- Biopulping—Paper is made from cellulose fibres, which must be separated from a tough wood fibre called lignin. The step by step process used to separate cellulose from lignin and other wood components is known as pulping. It is a time and energy consuming process, involving the mechanical processing of wood or the treatment of wood with harsh chemicals. In biopulping, cellulase and xylanase enzymes made by lignin-degrading fungi are used to pre-treat wood and break down the lignin fibres. Removing lignin prior to further wood pulping saves time and energy, and decreases the quantities of chemicals used.

- Draining—Enzymes can also improve water drainage during wood pulping, a process that often slows down paper production. When fine lignin fibres are degraded by enzymes, less water is absorbed, thereby reducing drainage times, lessening the energy required to dry the paper, and producing a cleaner water runoff.
- Bleach Boosting—Lignin fibres that remain in wood pulp are coloured and must be bleached, usually by harsh chlorine compounds under high pressures. As an alternative, enzymes may be used to remove fine surface fibres, thereby reducing the bleaching process or eliminating it altogether.

Enzyme Biotechnology and Textiles

Enzymes are used to treat and modify fibres, particularly during textile processing and in caring for textiles afterwards. For example, enzymes called catalases are used to treat cotton fibres and prepare them for the dyeing processes. Some bacterial enzymes are used to separate the tough stem of the flax plant from the flax fibres used in textiles. By degrading surface fibres, many enzymes, including some cellulases and xylanases, are used to finish fabrics, give jeans a stonewashed effect, or help in the tanning of leathers. A recombinant enzyme called laccase, made by certain fungi, may also be used to treat fabrics and even catalyse the synthesis of some synthetic fibres. There are even enzymes in regular laundry detergent to help break down dirt, clean clothes more effectively, and prevent the dulling of fabric colours.

Enzymes are frequently used in laundry detergents. Without them, very high temperatures and mechanical shaking would be required to effectively clean clothes and other textiles.

Enzyme Biotechnology and Biofuels

Enzymes may be used to help produce fuels from renewable sources of biomass. Such enzymes include cellulases, which convert cellulose fibres from feedstocks like corn into sugars. These sugars are subsequently fermented into ethanol by microorganisms. A new process called Simultaneous Saccharification and Fermentation has greatly improved ethanol production efficiency. In this new process, cellulase enzymes and fermentation microorganisms are combined in a single reaction mixture to produce ethanol in one step, rather than producing sugars from cellulose and then fermenting them into ethanol separately.

Advances in science and technology have allowed researchers to improve enzymes through modifications to enzyme producing microorganisms, or through direct changes to the enzymes themselves.

- Genetic Engineering and Recombinant DNA technology—By using recombinant DNA technology, microorganisms may be genetically modified to produce a desired enzyme under specific conditions. This is accomplished through recombinant DNA technology, whereby small circular pieces of DNA, known as plasmids, are used to insert enzyme-producing genes into the genomes of organisms that possess another desirable trait, such as the ability to thrive on inexpensive nutrients. Therefore, both the enzyme and the original trait will be expressed in a single recombinant microorganism.

- Protein Engineering—All enzymes are made of proteins, which are large molecules formed from basic units, called amino acids, strung together like beads on a chain. To form functional enzymes, long chains of amino acids must be folded properly. Some enzymes consist of only one chain, whereas others are made of several chains that fit together. Scientists are using technology to study how proteins are formed, how they fold, and how they function. By studying the relation between the structure of a protein and how it functions, they are developing ways to improve and engineer enzymes. Proteins may be modified by changing one or more amino acids, and/or changing the way the amino acid chains fold and fit together.

Enzymes are a sustainable alternative to the use of harsh chemicals in industry. Because enzymes work under moderate conditions, such as warm temperatures and neutral pH, they reduce energy consumption by eliminating the need to maintain extreme environments, as required by many chemically catalysed reactions. Reducing energy consumption leads to decreased greenhouse gas emissions by power stations. Enzymes also reduce water consumption and chemical waste production during manufacturing processes. Because enzymes react specifically and minimize the production of by-products, they offer minimal risk to humans, wildlife, and the environment. Enzymes are both economically and environmentally feasible because they are safely inactivated and create little or no waste; rather than being discarded, end-product enzymatic material may be treated and used as fertilizer for farmers' crops. Bottom of Form

(e) Microbial Enhanced Oil Recovery

Microbial Enhanced Oil Recovery (MEOR) is the use of microorganisms to retrieve additional oil from existing wells, thereby enhancing the petroleum production of an oil reservoir. In this technique, selected natural microorganisms are introduced into oil wells to produce harmless by-products, such as slippery natural substances or gases, all of which help propel oil out of the well. Because these processes help to mobilize the oil and facilitate oil flow, they allow a greater amount to be recovered from the well.

MEOR is used in the third phase of oil recovery from a well, known as tertiary oil recovery. Recovering oil usually requires two to three stages, which are briefly described as follows:

- *Stage 1: Primary Recovery* – 12% to 15% of the oil in the well is recovered without the need to introduce other substances into the well.
- *Stage 2: Secondary Recovery* – The oil well is flooded with water or other substances to drive out an additional 15% to 20% more oil from the well.
- *Stage 3: Tertiary Recovery* – This stage may be accomplished through several different methods, including MEOR, to additionally recover up to 11% more oil from the well.

The microorganisms used in MEOR can be applied to a single oil well or to an entire oil reservoir. They need certain conditions to survive, so nutrients and oxygen are often introduced into the well at the same time. MEOR also requires that water be present. Microorganisms grow between the oil and the well's rock surface to enhance oil recovery by the following methods:

- Reduction of oil viscosity – Oil is a thick fluid that is quite viscous, meaning that it does not flow easily. Microorganisms help break down the molecular structure of crude oil, making it more fluid and easier to recover from the well.
- Production of carbon dioxide gas – As a by-product of metabolism, microorganisms produce carbon dioxide gas. Over time, this gas accumulates and displaces the oil in the well, driving it up and out of the ground.
- Production of biomass – When microorganisms metabolize the nutrients they need for survival, they produce organic biomass as a by-product. This biomass accumulates between the oil and the rock surface of the well, physically displacing the oil and making it easier to recover from the well.
- Selective plugging – Some microorganisms secrete slimy substances called exopolysaccharides to protect themselves from drying out or falling prey to other organisms. This substance helps bacteria plug the pores found in the rocks of the well so that oil may move past rock surfaces more easily. Blocking rock pores to facilitate the movement of oil is known as selective plugging.
- Production of biosurfactants – Microorganisms produce slippery substances called surfactants as they breakdown oil. Because they are naturally produced by biological microorganisms, they are referred to as biosurfactants. Biosurfactants act like slippery detergents, helping the oil move more freely away from rocks and crevices so that it may travel more easily out of the well.

4

Agricultural Biotech-Derived Products: A Case Study

AGRICULTURAL BIOTECHNOLOGY-DERIVED PRODUCTS: GLOBAL TRACEABILITY AND LABELING REQUIREMENTS

The range of requirements instituted in other countries with respect to labeling and traceability for current plant-based products of modern agricultural biotechnology is broad. Although most countries have sanitary and phytosanitary regulatory requirements, some countries have no regulatory or labeling frameworks specific for products of modern biotechnology, some have evolving frameworks, and some have agricultural biotechnology-specific regulatory frameworks that may or may not require mandatory biotechnology-specific labeling. In the countries that have mandatory biotechnology-specific labeling systems, these requirements range from mandatory labeling for all products to mandatory labeling only for certain products or uses to labeling based on compositional differences.

Additionally, countries may incorporate a variety of rules or protocols under which particular products may be exempted from labeling requirements. At this time, only the European Union has instituted the double requirement for mandatory biotechnology-specific labeling and biotechnology-specific traceability. On the other hand, the United States neither requires that labels of food and feed products derived from genetically engineered plants indicate the method by which the plants were produced nor requires specific tracing of biotechnology products through the food and feed chain.

For regulatory frameworks without mandatory biotechnology-specific labeling requirements, products are sourced and delivered using existing commodity and finished-product markets without any special identification of transgenic products. In countries where there are mandatory biotechnology-specific labeling requirements, food manufacturers generally have chosen to avoid labeling their products based on such requirements. Supply chains that source non-biotech grain and grain products have been evolving to successfully avoid labeling or permit exemption of the products from labeling, depending on the particular requirements and exemptions established. With such supply chains involving specific grain sourcing and special programs to avoid commingling of sourced grain with bulk commodity grain, the fungibility of the product and the flexibility of the production systems are reduced and increased costs are absorbed somewhere along the supply chain. The greater the commercial requirements for documentation and segregation from the commodity chain, the greater the costs associated with originating and manufacturing particular food or feed ingredients. In addition, the greater the potential for pass-back of various liability claims back up the food and feed chain, the greater the potential impacts on costs and on the sustainability of production and delivery systems.

The EU's requirements for labeling and traceability for biotech grains and grain products, which includes the unique requirement for labeling irrespective of detectability have additional commercial implications for grain and grain product trade with the EU. Under the EU's regulatory framework, suppliers who do not wish to label are required to show through documentation that the grains and grain products are originated through identity preservation (IdP) systems and as such are exempt from labeling and traceability requirements. The documentation requirements and the challenge of supplying products that are exempt from labeling in the EU have significant impacts on the ability of U.S. producers to compete in the EU market. Two international fora, the Codex Alimentarius and the Cartagena Protocol on Biosafety (the Protocol or BSP), are also involved in the issue of biotechnology labeling and traceability. Codex, which establishes international quality and food safety standards, is currently having discussions on transgenic traceability and labeling. The BSP, which establishes requirements for the movement of "living modified organisms" (LMOs) between countries, is in the process of setting forth documentation requirements for commodity shipments of LMOs for food and feed purposes. The U.S. is not a Party to the Protocol. Requirements relating to documentation of the presence of LMOs in shipments of products in international commerce provide an additional challenge for U.S. exporters.

The following are other issues that are raised by the labeling and traceability requirements and the food and feed chain's efforts to implement them in practice.

- "Adventitious presence" or AP, defined in this case as the unintended presence of a small amount of transgenic material in seed, grain, or feed and food products, poses challenges both in terms of regulatory compliance and consumer/commercial expectations. The issue is complicated by the lack of applicable standards or guidelines and has led to some disruption of trade flows and trade disputes.

- IdP systems are increasingly important tools for preventing commingling of biotech and non-biotech grain and grain products and for documenting the handling and processing steps for particular products. Those systems can capture added value, manage regulated products, and segregate products with limited marketability, but they may also impose new costs and commercial risks.
- Tracing and testing are interrelated tools for origin or process certification, meeting customer traceability needs, safety verification, recalls, and inventory control. For many uses, tracing has evolved independent of transgenic product labeling requirements and different members of the supply chain use different approaches to address their specific needs. Traceability systems are evolving today from both ends of the supply chain to meet different market needs.
- Contracting requirements and any potential liability that might arise from biotechnology-specific labeling requirements vary depending on one's position in the food and feed chain and the type of product or ingredient involved. Contracts are typically used in all transactions along the chain except in consumer purchase of end products. However, contract terms relative to liability transfer with a transaction are not uniform along the chain.
- A variety of test methods have been developed to detect specific biotechnology-derived products, but no single test works for all food products or for all genetic events. There are as yet no internationally accepted standards for testing and the tests themselves can give variable results. Under some labeling systems, commercial risk exists because a

grain product grown from compliant seed, delivered as compliant grain to a processing facility, processed to produce a compliant grain product, and incorporated into the finished food may not test "in compliance" under some national compliance testing programs.

- When food manufacturers and ingredient suppliers comply with labeling regulations by sourcing ingredients that are exempt from labeling, contractual specifications with respect to IdP and tolerances can vary depending on the level of the manufacturers' commercial risk aversion. In some cases, the contractual specifications can be far greater than actual regulatory requirements.
- Increasingly stringent product requirements could lead to market segmentation, increased commercial risks, increased testing, and a narrowing of the group of suppliers. Product value becomes less transparent in the marketplace. These trends have different impacts on the commodity system, the food system, and in the political sphere.
- Consumers' interest in seeking additional information about their food is increased by variability in retail requirements in different countries and associated debates regarding safety, environmental concerns, and general acceptability of biotech crops.

I. INTRODUCTION

Charge

This report was prepared by the AC21 in response to a request from the Office of the Secretary of Agriculture to examine the current issue of the proliferation of traceability and mandatory labeling regimes for biotechnology-derived products in other countries, discuss the implications of those regimes, and describe what actions the food and feed supply chain is taking in response to those requirements for products shipped to those countries.

Process

To draft this report, a work group was formed after the AC21's third plenary session on December 4-5, 2003. The Traceability and Labeling Work Group benefited from participation by *ex officio* members of the AC21 and USDA staff. The Work Group members drew on both their own experience and expertise as well as information provided by outside experts, ex officio members of the AC21, and employees of USDA with relevant expertise. The Work Group met via conference call 8 times to discuss the mandate of the committee and draft a report that would address the Secretary's charge.

The working group activities and their written products were reviewed by the whole AC21 at the next 5 plenary meetings held between March, 2004 and February, 2005. The results of those discussions were incorporated into this final document. Thus, this report is submitted to the Secretary on behalf of all the members of the AC21 and is the product of the whole committee.

Scope

This report focuses on the impacts and implications of the various labeling and/or traceability frameworks currently in place globally. The wide range of frameworks reflects varying political, cultural and scientific perspectives. This report does not attempt to discuss the merits or values of the frameworks. This report addresses current plant-based food and feed products only. Grain and grain products, as the most widely traded genetically engineered crops, are used as the prime examples throughout this document. However, the implications of traceability and labeling requirements are the same for non-commodity or specialty food crops.

This report was developed in response to a specific request from then-Agriculture Secretary Ann Veneman for a summary limited to the current state of global traceability and labeling systems and their impacts on U.S. agriculture and food and feed industries. The report is being submitted along with a second report entitled, "Preparing for the Future," which is part of the AC21's continuing effort to address its main charge to examine how biotechnology is likely to affect agriculture and USDA over the course of the next 5 to 10 years.

The remainder of this report is divided into three parts. Section II consists of a description, in generic terms, of the range of systems that exist on a national level for labeling and traceability in other countries, as well as a description of current (and evolving) requirements under international agreements that impact trade in these agricultural products. That section, in its description of the range of current systems in other countries, also discusses how industry endeavors to meet the requirements for each. Section III delves further into commercial implications and realities around specific themes relevant to meeting traceability and labeling requirements, i.e., adventitious presence (AP), identity preservation (IdP) systems, the scope and range of traceability systems, contracting and liability, commercial risks associated with testing procedures and methodologies, the discrepancy between regulatory and true commercial requirements, and the impacts of market segmentation. Section IV is a discussion of policy concerns and issues that arise from this analysis.

Terms Used

Approved products are products containing or consisting of transgenic events that have passed the necessary regulatory requirements for food and/or feed use in a particular country or region.

Biotechnology is a range of tools, including traditional breeding techniques, that: (1) alter living organisms (or part of organisms) to make or modify products; (2) improve plants or animals; or (3) develop microorganisms for specific uses. This report will focus on a subset of products, namely those produced through genetic engineering or recombinant DNA processes, and the derivatives of those products. In this report, the terms "products of modern biotechnology" and "transgenic (or genetically engineered) organisms" (or their products), are used interchangeably to refer to these organisms. The terms, "genetically modified" or "genetic modification" (or "GM") are avoided except where they are part of a specific label indication.

Biotechnology-specific labeling refers to labeling that expressly indicates the means by which the product was produced, i.e., through the use of biotechnology, genetic engineering, or, according to some definitions, "genetic modification."

Commercial risk refers to the financial exposure and/or potential market losses associated with the sale, transport, shipment, or use of the grain and grain products resulting from contracts or other mechanisms.

Customers are individuals or groups who purchase, are provided with, or use, products and services.

Grain and grain products refers to wheat, corn, soybeans, etc., and products or ingredients derived from them, including finished foods.

Industry or the food and feed supply chain refers to all of the technology providers, seed breeders and producers, farmers, elevator operators, domestic and international shippers, grain processors, food manufacturers, distributors, wholesalers and retailers, that are involved in bringing food and feed products to the consumer.

Labeling refers both to information that appears on products offered to consumers as well as to documentation requirements accompanying shipments of grains, oilseeds, fruits, and vegetables, except where distinctions between the food labeling and shipping documentation are specifically noted.

Retail consumers are individuals who personally use or consume a good or service from a retail establishment.

Traceability refers to the ability to document a history of the origin, participants, steps and handling involved in the production of a food or feed product.

II. NATIONAL AND INTERNATIONAL REQUIREMENTS WITH RESPECT TO TRACEABILITY AND LABELING

As products of modern biotechnology have entered commerce internationally, over the past decade or so, countries have developed, or begun to develop, safety and identification requirements for new products, and in this process have reviewed the adequacy or applicability of their existing broader regimes for conventional product categories to the new products. Some countries have adapted existing regulatory systems and labeling and identification requirements for new products while other countries have developed a new system(s) for some or all of these products. While these actions have been taken for a variety of reasons, these additional requirements impose additional costs and risks for the food and feed supply chain. Such costs and legal and/or commercial risks derive both from difficulties in reliably meeting biotechnology-specific specifications for ingredients derived from commingled, commodity-derived raw materials and from concerns about the responses of consumers to products with biotechnology-specific labels. The magnitude of the costs and commercial risks depends on the particular requirements in place.

The United States government, which does not require mandatory biotechnology-specific labeling for transgenic products, has argued in numerous international fora, especially the Committee on Food Labeling under the Codex Alimentarius, that there is no scientific justification for the imposition of mandatory labeling requirements based solely on the process by which the product was produced. At the same time, however, USDA has attempted to help markets meet customer demands around biotechnology content through the provision of tools and services relating to product testing and evaluation of testing services, process verification, and technical information for grain and seeds.

This report does not address the merits or implications of particular national regulatory systems or any requirements therein in the evaluation of food or environmental safety for transgenic products.

A. Regulatory Framework Categories and Compliance Implications

A host of agricultural products move between countries through international trade. For products of modern biotechnology there may be regulatory compliance implications related to trade in these products depending on the status of the event approval or the decision of the food manufacturer regarding labeling of the final food product in a specific country. For this discussion, the section will focus on the commercial impacts of the various regulatory frameworks on supplying "approved" transgenic products.

There is a wide range of frameworks for transgenic products depending on both national and international requirements associated with trade with or in a particular country. While, at the most basic level, frameworks can be divided into those requiring biotechnology-specific labeling and/or traceability and those without such requirements, in practice, there are significant distinctions to be drawn among subcategories within these two. The nature of the overall regulatory framework for safety evaluation of agricultural biotechnology products is also relevant. By examining the spectrum of labeling and traceability requirements it is apparent that compliance in these markets is complex and in many cases requires additional processes and procedures and generates increased cost to supply grain and grain products.

The frameworks can be divided as follows:

1. Regulatory or Labeling Frameworks that have no Specific Provisions for Products of Modern Biotechnology

Most countries have not considered or adopted regulatory frameworks specific for agricultural biotechnology. Such countries have no biotechnology-specific regulatory restrictions on grain and grain products that are generally sourced and delivered using the established commodity systems for grain. Established phytosanitary procedures for the movement of grain apply for such products as they do for all grain and grain products. Grain and grain products are bought and sold based on national specifications for grain and grain products through buyer and seller contracts and in some cases international mechanisms. Participants in these markets capture the values associated with economies of scale and importers are able to maintain low costs based on flexibility and competition.

2. Regulatory Frameworks Specific for Agricultural Biotechnology Products that do Not Require Mandatory Biotechnology-Specific Labeling

A number of countries have adopted specific regulatory frameworks for agricultural biotechnology. For several of these countries, after approval, transgenic products are generally considered equivalent to conventional products and as such do not require specific labeling in the marketplace. After approval, there are no restrictions on these grain and grain products, and consequently these products are sourced and delivered using the established commodity systems. Grain and grain products containing approved transgenics are bought and sold based on national or customer specifications through buyer and seller contracts and in some cases international mechanisms.

Participants in these markets capture the values associated with economies of scale and importers are able to maintain low costs based on flexibility and competition. In cases where a food manufacturer may consider there to be a marketing advantage associated with labeling a product as "non-GM" or "not made from transgenic grain," there may be a specification of identity preservation and/or sourcing from non-biotech grains. In this case, the manufacturer can label the food voluntarily in the above ways to differentiate the product in the marketplace.

These additional requirements to serve this market, such as tolerances for AP or IdP requirements or expectations, are contractually negotiated and the added costs are priced according to market forces and availability of the product. (Part III of this report contains further discussions on AP and IdP.) The additional costs to originate and manufacture these products are sometimes passed along directly to the consumers.

3. Regulatory Frameworks Specific for Agricultural Biotechnology that Require Mandatory Biotechnology-Specific Labeling.

Several countries have adopted specific regulatory frameworks related to agricultural biotechnology that involve mandatory labeling for foods containing ingredients derived from transgenic products. These requirements are most often independent of the product risk assessment and often based on a "right to know" policy. The details of when such labeling is required and when products can be exempted from requirements are usually decided nationally. There are several compliance scenarios for grain/grain products based on national legislation and how national and local regulatory officials interpret that legislation. While there are multiple permutations, there are a number of different overall scopes for regulatory frameworks for labeling in the marketplace.

(a) Mandatory Biotechnology-Specific Labeling Frameworks: Scope

1. Mandatory Biotechnology-Specific Labeling of all Products.

In some countries, all grains and grain products that fall within the scope of a national definition of a transgenic or "genetically modified" organism are subject to mandatory transgenic product labeling regulation. In this case, if the grain or grain product is originated from a country that has commercialized a transgenic variety, then all products must be labeled unless they qualify for an exemption.

This framework places a significant number of grain and grain products within the scope of the regulation and can have significant impacts on the supply chain for grain and grain products. In general, food manufacturers have chosen to avoid labeling their products and, accordingly, in countries with mandatory labeling of all products, the development of supply chains to avoid or permit exemption of the products from labeling has been proceeding.

These supply chains usually require two basic specifications, IdP and a tolerance for AP. The requirements to source grain or grain products from non-biotech grain sources and to adopt specialty programs and systems to avoid commingling with commodity grains reduce fungibility and flexibility and increase cost. The greater the requirements for documentation and segregation from the commodity stream, the greater the costs associated with originating and manufacturing these ingredients. Most requirements for IdP allow for a low level of accidental commingling. National requirements regarding the allowed level for AP often differ because the acceptable level of commingling is not dictated by scientific risk assessment. In many regulatory schemes, the allowed AP level may in effect be based on the level of testing detectability or on perception. In general, the lower the level of AP required, the greater the associated costs.

2. Biotechnology Specific Labeling for Specific Products and Uses.

In some cases, only specific grain and/or grain products fall within the scope of labeling. In general, a country's regulatory authority will publish a listing of specific grain and/or grain products that are "within the scope of the mandatory labeling regulation." Depending on the types of products listed, there may be significant impacts on how food manufacturers source these ingredients.

Specific product listing can be applied not only to ingredients, but also to specific end-use products. For example, based on the current Chinese regulatory framework, soybean oil derived from transgenic soybeans, bottled and used as a salad dressing, must be labeled as "soybean oil-genetically modified." However, soybean oil used as an ingredient in a food product does not have to be labeled in the final food.

In general, food manufacturers to date have chosen to focus on producing only products for which a biotechnology-specific label is not required, because of the concern that consumers could interpret a "biotech label" as a warning and this could threaten hard-won trust in their established brands. Therefore, in countries with biotechnology-specific product and use requirements, products are typically produced using specialty supply chains that enable manufacturers to avoid such labeling. However, food manufacturers and other elements in the food chain have also expressed practical concern about the whole approach due to the difficulty in reliably sourcing ingredients that would not trigger labeling requirements.

3. Labeling Requirements Triggered when Modifications Result in Compositional Novelty

In this case, transgenic-derived products in which the composition or functional properties of the grain or grain product have been modified such that they are no longer substantially equivalent to their conventional counterpart require the mandatory labeling of the compositional change. In some countries, like Japan, compositional novelty also triggers mandatory biotechnology-specific labeling.

In practice, there are presently very few products that trigger this labeling requirement. There are no international standards that allow the regulator to determine if the product is outside of the range of substantial equivalence, although there are a number of initiatives to define this at an international level. In the case where the new product mimics a conventionally bred product, it may or may not be exempt from labeling depending on the country.

As more products come to market with modified functional/compositional characteristics, this labeling requirement may become commercially significant. It would not be unrealistic to anticipate that food manufacturers will look to avoid labeling of food ingredients that would fall within the scope of this mandatory labeling requirement. It is worth noting, however, that products may evolve with consumer benefits resulting from such novelty, and in such instances, manufacturers wish to inform consumers, in order to enable consumer-based demand.

4. Evolving Regulatory Frameworks

In this case, the importing country is in the process of implementing a regulatory framework and the requirements for compliance are in transition and can be unclear. In some instances, regulatory requirements may be suspended during a transitional period; in others, regulations may be enforced immediately.

Between the initial enactment of labeling requirements and the time the biotech events in the "queue" have been authorized, there is generally a period of time where either no events are approved for import, or only some events are approved for import. Availability or lack of transitional measures can have significant commercial implications for food company compliance.

In some cases, the implementation and enforcement of these new regulations are variable and expanding in scope. These markets require a high degree of diligence and attention during this period. The availability of a transparent and predictable regulatory framework can enable international trade to continue without significant disruptions.

(b) Mandatory Biotechnology-Specific Labeling Requirements: Exemptions

When a grain or grain product is within the scope of

mandatory labeling requirements, most food manufacturers try to originate grain or grain products that do not require a biotechnology-specific label. The conditions under which products or ingredients are exempt from labeling requirements are determined nationally and supply chain requirements are diverse. These conditions can significantly impact the capability to supply and the costs associated with these ingredients.

1. Exemption from Labeling Requirements Based on Following Identity Preservation Protocols and Meeting Specific Tolerances

This is the most common exemption from mandatory labeling. This exemption allows food manufacturers to originate grain and grain products from non-biotech grain through specialty programs in order to avoid process-specific labeling. It requires both the development of supply chains that are parallel to the commodity system and strict separation of these specialty grains from the commodity system.

In this case, the ingredient is exempt from labeling if it was originated from non-biotech grain and steps have been taken to avoid commingling with commodity grain products. The level of detail required in supply chain documentation is often dictated by national regulation and may be specific to country or product use. Guidance documents are sometimes available. In general, it is accepted that, in practice, IdP of non-biotech grain cannot completely exclude transgenic-derived grain. Consequently, some governments have assigned a tolerance level for the AP of biotech grains in a shipment of grain that is exempt from labeling. Tolerances set for AP are often country-specific and may be based on a variety of factors, including detectability, test method performance, marketplace, consumer, and political considerations.

The greater the requirements for documentation and segregation from the commodity chain, the greater the costs associated with originating and manufacturing these ingredients. Costs are also related to the degree and nature of regulatory requirements and enforcement. In general, the lower the level of AP allowed, the greater the associated costs. Commercially, the cost and the degree of difficulty to service these markets depends on the ability to source non-biotech grain, the ability to access systems to supply this grain to specific markets, the relative penetration of the biotech crop in a region, and the locations where transgenic grain types have been planted. In addition, the potential for the extension of liability related to either human, animal or plant health or environmental impact can affect the cost structure as well as the sustainability of production and delivery systems.

2. Exemption from Requirements based on Status of Ingredient

Under these regulatory frameworks, only the major ingredients are subject to mandatory labeling and minor ingredients are exempt from labeling if they represent less than a certain proportion of the final food. For example, an ingredient might be exempt from labeling if it is not one of the top three major ingredients.

Whether or not the grain or grain products are major ingredients in the final food can therefore be significant for the food ingredient supplier. For major ingredients, the vast majority of food manufacturers are avoiding labeling through IdP and compliance with tolerance requirements. For minor ingredients, food manufacturers can source ingredients from commodity sources.

3. Exemption from Requirements based on Detectability

This is a very common exemption in countries with mandatory labeling. Under this regulatory framework, a grain product derived from a product of modern biotechnology is exempt from labeling if it no longer contains detectable protein or DNA based on available detection methods.

Several food ingredients are derived from refining processes that remove the DNA and protein during processing. Such products can be produced from commodity grains and used in foods without labeling the final product as derived from modern biotechnology. This exemption has significant commercial impacts as it exempts highly refined products from such labeling.

The only countries currently without this exemption are the members of the European Union (EU), Brazil, and China.

4. Regulatory frameworks that require mandatory biotechnology-specific labeling and traceability.

The requirement for mandatory biotechnology-specific labeling and traceability is unique to the European Union. "Traceability" in this instance refers to the ability to trace the transgenic events in the grain used to manufacture the food ingredient. All grain and grain products that are derived from commodity grains for which a transgenic variety has been commercialized in the producing country would be subject to EU labeling and traceability requirements, based on the fact that a grain or grain product was not intentionally sourced from an IdP system and therefore likely contains transgenic material. As such, the term "traceability" is specific to products that are within the scope of mandatory labeling requirements and would need to be labeled in the marketplace. The term "traceability" here does not apply to products exempt from labeling, as such products derived from conventional material do not have transgenic events to trace through the supply chain. In the absence of precise guidance from EU regulatory authorities, several different interpretations of this requirement are emerging in the marketplace.

The EU's requirements for labeling and traceability for biotech grains and grain products, which includes the unique requirement for labeling irrespective of detectability have additional commercial implications for grain and grain product trade with the EU. Under the EU's regulatory framework, suppliers who do not wish to label are required to show through documentation that the grains and grain products are originated through IdP systems and as such are exempt from labeling and traceability requirements. The documentation requirements and the challenge of supplying these niche products will have significant impacts on cost and ability to participate in this market. Based on the current infrastructure, originating these materials in large volumes may be difficult and expensive. In addition, compliance and enforcement mechanisms in the EU and its Member States for these new requirements are as yet uncertain.

B. International Requirements

The issues of labeling and traceability have been introduced to some extent at the international level in two forums: Codex Alimentarius and the Cartagena Protocol on Biosafety. In addition, agreements under the World Trade Organization framework and the International Plant Protection Convention bear on the application of traceability and labeling regulations.

1. Codex Alimentarius Commission

Under the auspices of the United Nations Food and Agriculture Organization (FAO) and the World Health Organization (WHO), the Codex Alimentarius Commission was established to develop internationally acceptable standards for use in the area of food quality and food safety. The Codex Alimentarius is significant in international food trade in that both the WTO Agreement on the Application of Sanitary and Phytosanitary Measures (the SPS Agreement) and the WTO Agreement on Technical Barriers to Trade (the TBT Agreement) encourage the international harmonization of food standards. In addition, the SPS Agreement specifically cites Codex standards, guidelines and recommendations as the preferred international measures for facilitating international trade in food.

The Codex Commission has established subsidiary bodies to develop standards, guidelines, and recommendations. These subsidiary bodies (committees, ad hoc groups, etc) have participation from Member governments. In addition, international intergovernmental organizations, as well as representatives from international nongovernmental groups, attend as "observers". To initiate the development of a standard, proposals are brought forward by a national government or a subsidiary committee of the Commission. Once it is adopted as a draft standard, the proposal undergoes a stepwise process to achieve approval by national governments. It can take several years before the process is completed.

The issue of "traceability" has been introduced in several Commission subsidiary bodies (Committee on General Principles, Committee on Food Import and Export Inspection and Certification Systems, Committee on Food Labeling) with no major resolutions to date. One of the major issues is whether traceability requirements for food are appropriate specifically for the subset of food products that have been derived from modern biotechnology. The countries of the EU, having national legislation requiring traceability of 'GMOs', support broad definitions in Codex in order to advance their national requirements internationally. The U.S. maintains that those definitions are too broad and not based on existing Codex principles related to safety or quality of products. The subsidiary body established to deal specifically with biotechnology issues (Ad Hoc Intergovernmental Task Force on Food Derived from Biotechnology) has identified several specific potential applications for product tracing generally and recognized that product tracing can be applied outside of biotechnology applications. The Task Force has noted that these applications should be consistent with the provisions of the SPS and TBT Agreements and that application of tracing under these agreements is under consideration within Codex under its Executive Committee.

The issue of labeling of products derived from biotechnology has been introduced in Codex and discussions have dealt with labeling of products derived through biotechnology for food and food ingredients, feed and feed ingredients, and, most recently, prepackaged foods. Discussions on labeling have centered on consumer "right-to-know" rationales for labeling versus the desire to label only for safety related issues, and have reached no conclusions.

Discussions under Codex on these matters continue to evolve. If biotechnology-specific labeling does further develop in the Codex forum, issues such as development of thresholds, detection methods, lists for inclusions and exclusions, labeling language, and supporting documentation requirements, etc. will likely come under discussion. Given the number and complexity of these issues, it may be difficult to come to an international agreement that can be effectively used in international food trade.

2. Cartagena Protocol on Biosafety

The Cartagena Protocol on Biosafety (the Protocol, or BSP), developed under the auspices of the Convention on Biological Diversity, was adopted in January 2000 and went into force in late 2003. A main focus of the Protocol is the establishment of an advance informed agreement (AIA) procedure to ensure countries are provided necessary information to make informed decisions before importation of a "living modified organism" (LMO) for intentional release. AIA is not required for LMOs intended for food, feed, or for processing. The U.S. is not a Party to the Convention and thus cannot be a Party to the Protocol. The Protocol also allows for Parties and non-Parties to enter into bilateral and multilateral agreements to implement requirements of the Protocol, consistent with the Protocol's objectives.

The Protocol defines requirements for documentation accompanying "transboundary" shipments of LMOs. The requirements are differentiated on the basis of use (destined for: contained use; release into the environment; or direct use as food or feed, or for processing) of the material. These documentation requirements are not labeling requirements for finished food products. The most controversial requirements are those for (typically commingled) commodity shipments that may contain one or many types of LMOs in a single shipment. The text of the Protocol allows the exporter to document that such commodity shipments "may contain" LMOs and are not intended for intentional introduction into the environment. However, the final language accompanying commodity shipments containing LMOs is still under discussion and must be completed by September 10, 2005. Major commodity exporting countries are in favor of retaining the "may contain" language as acceptable documentation, taking into account the practicalities of trade, while other countries desire more specific information (such as, most problematically for trade, identification of the specific LMOs in the cargo). Other specific aspects relating to documentation are also still under discussion.

The documentation requirements that will be established

under the Protocol for shipments of LMOs from a country (Party or non-Party) to a Party, will play a major role in dictating how easy or difficult it will be for exporters to comply with Protocol requirements, depending on how much information will be required on documents accompanying shipments. The U.S., Mexico, and Canada recently developed an arrangement to address documentation needs for commodity shipments that was intended to be consistent with the objectives of the Protocol. This arrangement allows for the continued use of the "may contain" language for affected commodity shipments in trade. It also establishes that a shipment that has 95 percent non-LMO content does not require the 'may contain' language, provided that this does not conflict with the regulations of the importing country. Of particular importance to commodity trade is the stipulation that, under the arrangement, adventitious presence should not be considered to be a trigger for the LMO documentation requirements. The extent to which the arrangement could serve as a model to facilitate trade between Parties and non-Parties will likely be influenced by the success of its implementation in North America and the strength of commodity exporting countries in the continuing negotiations around implementation of the Protocol.

3. International Plant Protection Convention (IPPC)

The IPPC affirms that contracting Parties have sovereign authority to regulate, in accordance with applicable international agreements, the entry of plants, plant products, and other regulated articles, and may, among other things, prescribe and adopt phytosanitary measures concerning their importation, provided such measures are technically justified, transparent and not applied in a way that constitutes either a means of arbitrary or unjustified discrimination or a disguised trade barrier. To the extent that any traceability or labeling requirement is part of a phytosanitary measure (e.g., for surveillance or pest reporting) the provisions of the IPPC may be relevant. The WTO recognizes the IPPC as the standard-setting body for phytosanitary measures.

4. World Trade Organization (WTO)

The WTO is not prescriptive with regard to labeling/ tracing requirements. The Technical Barriers to Trade (TBT) Agreement of the WTO requires that countries' technical regulations may not be more trade-restrictive than necessary to fulfill a legitimate objective. For many countries, consumers' "right to know" is considered to be a legitimate objective; thus, the TBT does allow for mandatory labeling and appropriate enforcement mechanisms, which could include some types of product tracing. However, the ability of all the various labeling and traceability regimes to meet the "no more trade-restrictive than necessary" requirement is not clear.

III. COMMERCIAL IMPACTS/REALITIES

There are several specific parameters that affect the U.S. industry's ability, in practical terms, to meet marketplace or regulatory requirements for process-specific labeling and/or traceability. This section describes some of the issues involved, some of the tools evolving to meet market demands (and the limitations of those tools), and some of the overall impacts of the changes that are being seen. The topics are not provided in any specific order.

A. Adventitious Presence

The term "adventitious presence" or AP refers to low levels of unintended material in seed, grain, or feed and food products. The key components of the definition are "low levels" and 'unintended."

AP of materials in the food and feed supply chain has always existed, based on the fact that different materials often share either infrastructure, space, or both. It also occurs due to the biology of plants, which allow different varieties to sexually reproduce with one another. While AP can be minimized, as a practical matter it cannot be eliminated entirely and is not unique to crops enhanced through biotechnology. To deal with AP of different products, allowances for the unwanted material have been recognized in laws, regulations and standards that establish allowances for these materials. In the United States, under the Federal Seed Act, for example, 5 percent of seed corn is allowed to be off-types and still be labeled as a specific corn seed hybrid. Allowances for AP also exist in grain and foods. With respect to grain, USDA has set grade standards for foreign material.

Some amount of AP is to a large extent unavoidable in grain and grain products based on the realities of plant biology, seed production, and the distribution of commodity crops. There are a number of factors that contribute to commingling of grains. Pollen flow, volunteerism, mixing during harvesting, transport, storage and processing, human error, and accidents can all play a role in AP. In practice, material intentionally introduced into seed, grain or feed and food products does not meet the definition of AP.

AP for biotech, then, refers to unintended, low levels of transgenic material (or specific transgenic events) in seed, grain or feed and food products. AP can occur in four different ways creating different regulatory or commercial implications.

1. Events that have Met Regulatory Requirements in the U.S. and the Importing Country

The AP of approved events is an issue for trade outside

the United States as it impacts when a product may need to be labeled as containing transgenic material or is unwanted. In this case, there are no food or environment safety concerns and no legal or regulatory problems exist with the presence of these products. For food manufacturers who wish to avoid labeling (through tolerance exemption) or choose to exclude it for marketing reasons and for consumers who wish to avoid foods containing biotechnology-derived ingredients, AP of the biotech event can raise commercial and marketing issues.

For markets where companies wish to avoid labeling of products (exemption from mandatory labeling), the requirement to source products below a specific tolerance is likely to increase the costs and exposure to liability and risks associated with compliance and non-conformance. In addition, there are no international standards for allowable trace amounts of biotech products in non-biotech products and this has created a very complex marketplace. Commercial requirements can add another layer of complexity. For example, some commercial market requirements for AP at times exceed the requirements of government regulations.

In some cases, companies wish to exclude biotech materials for marketing reasons and may or may not decide to claim it on the package (voluntary labeling). In the case where complete avoidance is required, AP becomes a difficult issue to manage. A primary driver of this commercial practice is consumer expectation with respect to AP in the product. It is expected that some customers want zero tolerance for AP. In practice, food manufacturers require some tolerance for AP and as such struggle with what constitutes an "acceptable zero". This has yet to be defined. Moreover, it is not clear who will ultimately define "acceptable zero" on a commercial basis.

2. Events that have Met Regulatory Requirements in the U.S. But Not Elsewhere

Adventitious presence can involve material that has completed the federal regulatory process in the U.S. and has been commercialized domestically but has not yet met regulatory requirements in countries that may import the grain or grain products. Once a biotech product is widely commercialized in a production area and handled through the same handling systems as commodity grains, the ability to supply grain with zero risk of adventitious presence is not possible. The adventitious presence of this type of biotech event raises legal, commercial and marketing issues because without regulatory approval in the importing country, there is generally a zero tolerance policy for the presence of these events in a grain or grain product.

Given the biology of plants, and the increasing ability to detect ever more minute traces of genetic material, zero tolerance is in practice an unachievable standard. Concerns about the commercial implications of AP of this kind may result in alteration of the geographic patterns of raw material sourcing. Contracts often will require modification and new controls such as IdP systems may need to be implemented. The location, number and scale of manufacturing facilities may be affected. Each change has cost implications.

For events that have met regulatory requirements in the U.S. but not in importing countries, AP can result in blockage in the trade of seed, grain or feed and food to those countries.

3. Events that have Met Regulatory Requirements Elsewhere But Not in the U.S.

In this case, the issues are comparable to those for events that have met regulatory requirements in the U.S. but not elsewhere. While there are few examples of this at the time of writing this report, the US government is working to further address this issue. Currently, USDA and the Environmental Protection Agency (EPA) maintain a zero tolerance specification for events that have not yet completed federal regulatory requirements.

4. Events that have Not Met Regulatory Requirements in the U.S. or Elsewhere

AP, involving biotech material that has not gone through the federal regulatory process or that has not successfully completed the federal regulatory process, can occur, ending up in seed, grain or food that is marketed to customers. In this case, presence of this biotech material in the food or feed supply chain would create significant commercial disruptions and likely impacts on international trade. The presence of such biotech material in food or feeds and these products could be considered illegal by the U.S. government and could require removal from the marketplace. Like government policy in other countries, the United States' policy on AP of transgenic plant material intended for food use that has not yet completed U.S. regulatory reviews is evolving.

In general, the policy and commercial issues around AP in the context of plant biotechnology internationally relate to unapproved and approved transgenic events, to inconsistencies in applicable rules and regulations and their enforcement, and in the range of differing commercial market expectations. The consequences of AP in different contexts create both regulatory issues and commercial or contractual ones.

AP requires governmental and commercial responses. As a consequence of AP, governments and industry must deal with disruption of trade flows, including product recalls and shipment rejections, which may lead to trade disputes and WTO cases. The current uncertainty and costs associated with AP in general will continue to be significant issues until workable policies are instituted and common understandings reached among countries and in the marketplace. Special arrangements, such as the one developed by the U.S., Mexico and Canada to facilitate movement of biotech and non-biotech grain in a manner intended to be consistent with the Cartagena Protocol on Biosafety, may be devised to attempt to deal with legal requirements and AP.

Internationally, the issues around AP are complicated by the involvement of government and international organizations and treaties and the lack of standards or guidelines. As was described above, labeling and traceability requirements vary among countries, and can be either voluntary or mandatory. Regulatory policy development and enforcement mechanisms also vary widely. Creating greater regulatory consistency or harmonization of AP requirements and compliance by the U.S. government may improve the current impacts of AP in the marketplace.

B. Identity Preservation Systems

Identity Preservation (IdP) is the process of managing a product so that: 1) the product does not become unintentionally commingled with unlike product; 2) the handling and processing steps in producing the product are documented; and 3) intrinsic or management-related product characteristics may be capitalized upon or marketed. IdP relies on documentation, communication, and coordination of all parties involved in the process, and can also include external verification or certification parties.

In the context of biotechnology, IdP is used in agriculture today to segregate biotech products from non-biotech products or particular varieties with specific biotech traits from other varieties. IdP can be used to accomplish one or several goals:

Capturing Added Value. Added-value products are those that have a real (tangible or functional) value for a specific purpose or products that have a perceived value developed by meeting particular customer specifications. The value may come from specific properties intrinsic to the end product or from specific procedures followed in its production. Sometimes the value derives from the producer's ability to provide consistency in the delivered product, but it also may derive from verification of product origin or the production practices used to manufacture the product. To maintain value in these systems and protect the integrity of an added value product, all steps in the value chain typically must be explicitly identified. This is usually executed through the use of contracts and Standard Operating Procedures (SOPs) for production, storage, and delivery of the product. Depending on the system, the end user may supply the seed but generally the producer is provided guidance on the expectations required to meet the system specifications. The complexity of the SOP is related to the value of the product and the level of purity required by the end user. Generally the lower the tolerance or the greater the complexity, the higher the costs are for supplying such a product to the end user. Marketing usually occurs through a contractual arrangement between the producer and an end user.

Some producer groups have been formed to pool resources and talents attempting to create and capture added value opportunities for the production of specific products that meet specific market segment needs. Value can be created where premium prices are received, market share is captured, or productivity gains arise somewhere in the chain through preserving identity and possibly end product labeling.

Several traditional agricultural organizations have established producer certification systems to establish a higher level of education and certify that they are equipped to operate under various SOP or IdP requirements. A "Cost of IdP Calculator" is available on the internet to assist growers in determining the added cost of various contract opportunities. The producer then can predict whether the added value will offset the additional cost incurred to meet the specifications required by the end user or regulatory requirements required to avoid labeling.

A number of grain handlers, processors, and food companies have established proprietary programs to meet particular markets and specific customer needs. They have their own SOPs and pay variable premiums depending on the value in manufacturing or in the ability to capture higher value at the customer level. Since the main focus is on creating or capturing value through enhanced functionality or meeting consumer preferences, the market tends to allow the most efficient methods to evolve, tolerances tend to vary, and testing requirements are tailored to each situation. As such, IdP systems and requirements vary across geography, among importing countries, and all the way to specific end-use customer requirements. No system or set of requirements fits all circumstances.

Controlling Products Not Approved for Food or Feed Uses. These products often require containment to prevent the adventitious commingling of these products in food or feed. Under specific regulatory systems, some regulated articles may have limited approval but may be eventually intended for traditional uses. Others, such as those intended for pharmaceutical or industrial enzyme production, may never be intended for food and feed use. For these products, identity preservation systems are integrated into SOPs to contain the product as well as to preserve the integrity of the product. Producers are subject to training and certification. To date, these non-traditional products have been produced on very small acreages and many of these non-traditional uses for crops will remain niche products in small highly contained spaces. The opportunity may be only for a few producers. Isolation and the incorporation of several redundant containment systems are becoming, or have become, standard requirements, and the systems and SOPs are continuing to evolve toward even greater rigor and additional oversight, including both physical and biological containment.

Controlling Limited Marketability Products. Products with limited marketability in this context are those that have full food and feed approval in the U.S. but not in other export destinations where grain, ingredients, or finished products are sold and marketed. In most cases, this is an outcome of asynchronous approval processes or time lags in the approval processes for some countries. This is challenging to the supply chain because regulatory frameworks often have zero tolerance for these products even though they have been approved for use in other countries.

To deal with these export markets at the front of the supply chain, the seed industry and grower associations have developed informational systems to attempt to channel these products in markets where they are approved and attempt to exclude them from markets where they are not yet approved. For example, the Market Choices® certification program was developed by the seed industry to provide a distinctive symbol for producers to identify which corn hybrids possess traits not yet approved for export to the European Union and should be marketed in specific channels.

"Know Before You Grow" and "Know Where To Go" programs have been developed to better inform producers before they buy their seed and before harvest as to who is willing to accept and channel specific transgenic events in specific regions of the U.S. transgenic products.ronmental safety r implications of particular systems or requirements

For these types of products, IdP can start at the farm level with relationships, contracts and/or delivery mechanisms at the next stage of the supply chain (grain handling). Different companies have different positions on the acceptance and/or channeling of products with limited marketability. In addition, specific facilities may have restrictions or create opportunities for producers based on specific market forces that apply to their geographical area or the markets to which they ship product, whether for export or domestic consumption. Handling of limited marketability events continues to evolve in the marketplace but can be in conflict with high-throughput handling systems that are built solely for speed and rely heavily on blending or commingling to meet quality specifications. IdP is becoming more prevalent as new products and new market opportunities arise. In all cases, there are new requirements to execute these systems. In developing and utilizing these systems, the supply chain needs to make business decisions to incorporate the requirements for channeling versus segregation, the need for certification or third party verification, the impacts and commercial risks associated with testing tolerances, and the appropriate documentation to address compliance with market demands. With different supply infrastructures, expertise, and demands, several different systems are likely to evolve in the marketplace.

With new technology on the farm and throughout the food chain, the moving target of customer desires and market requirements will cause demand shifts in identity preservation systems as well. American agriculture and the U.S. food and feed chain have a strong history as an adaptable, market-focused integrated production system. This system is becoming better equipped and evolving towards meeting both the current and future opportunities and demands in the marketplace, but the challenges posed are greater for some chain segments than for others.

C. Traceability Systems

The following are some comments about tracing practices, many of which do not involve biotechnology, that already exist and/or are evolving in the marketplace. These systems, however, may be applied for biotechnology products. The principles of tracing and the tracing systems set forth below could apply for any specific reason, including products of biotechnology, for following grain through or within various levels in the supply chain. Testing is added as appropriate.

Tracing and testing are two interrelated mechanisms for understanding product identity and flow, verifying identity, or addressing safety concerns. Tracing provides a trail and testing may be used to demonstrate or underscore compliance with prescribed or agreed parameters.

The type of tracing employed varies broadly depending on its purpose:

- Origin or process certification—Tracing is accomplished through a paper trail with almost no testing and relies on self-declaration by all parties. The process is certified, usually on an annual basis, by a third-party certifying agency. Blending occurs with most products as they move through the supply chain. Tracing focuses on certifying process or attributes, such as organic, rather than on tracing to a specific location.
- Meeting the traceability needs of customers—Tracing

here relies primarily on layered spreadsheets built on units of delivery, i.e., the assembly of the cumulative record describing, as precisely as available data allow, what was delivered at each step in the value chain. The main concern is usually from where, as opposed to from whom, a product is coming. Testing is usually involved to verify the absence or presence of a specified event and may or may not equate with liability transfer if the event shows up later in the value chain. For export grain, for example, tracing would provide information that a particular vessel came from a specified port and contained a blend of grain from specified barges or railcars that came from specified elevators who bought from an identified group of farmers whose grain was again blended together through the elevator system.

- Safety verification – Safety verification, e.g., for mycotoxins, focuses primarily on testing of products and of ingredient streams (at various places). The amount of testing generally increases closer to the consumer in the value chain.
- Recalls – Tracing for recalls typically starts at the point in manufacturing or processing at which packaging takes place. In the U.S., this process is built on "one step forward one step back" model. Some product identifier, usually a bar code, allows tracing back in steps to the manufacturer, revealing when and where the product was made and what ingredients were used. Internal manufacturer records enable the determination of the source of ingredients and the narrowing of the possible time of manufacturing to a specified run, often within a few minutes. If the cause of a recall was a problem with an ingredient from a specific supplier, liability

for all costs would generally pass to that supplier. The supplier may or may not be able to trace the problem or pass the liability further up the supply chain, depending on the strength of its "one step forward, one step back" capabilities.

- Inventory control – Some systems utilizing newer technology are evolving in consumer-packaged goods and retail industries. Industry collaborations will allow tracking of inventories all the way from the manufacturing process from whatever point the newer bar codes or identifiers are placed on the product. Costs are relatively high to set up such a system (perhaps as much as $400,000 for an individual food warehouse). These structures could provide the basis for electronic product codes with "smart chips" in the future if they prove economically feasible and are accepted by consumers.

If all players in the value chain contribute to a "one step forward one step back" linkage, tracing could provide a trail through all the individuals or groups of individuals involved in blending components in ingredient streams.

The seed industry operates today by utilizing a highly documented trail governed by Federal and state seed laws and regulations from state certification organizations, including crop improvement associations. Every lot of seed is catalogued by variety through all stages from breeder to foundation to commercial seed. The respective parties in that chain keep all records of production, field inspections, inventories, transportation moves, cleaning, certification tags, etc. for a minimum of three years. Some movement toward the use of bar codes or electronic scanning is occurring. Spot inspections by state or crop improvement association personnel are conducted to verify compliance with regulations or certification requirements.

Currently, the commodity grain industry most often provides tracing by generating an identifying ticket, such as a scale ticket, at the point at which grain is first unloaded from the farm. Grain is primarily commingled; any partition of the commingled material is usually by quality or variety. Blending occurs to meet specifications of the sale contract. A shipment will be associated with an identifying number such as a bill of lading, railcar or barge number, container number, or vessel name. Testing is common at all points of transfer to verify quality, safety, or the absence or presence of a particular event or trait. It is not feasible to trace back to individual farmers, but tracing back today through the system of records to a group of elevators or farmers remains possible. All records from all steps are not on a common spreadsheet to allow instant recovery. However, a composite record probably could be reconstructed in a matter of hours or few days. Liability is usually passed with title verified by testing. If an undetected event showed up later, it is conceivable that there would be efforts to pass liability back up the chain, step by step, to the point where blending might render it impossible to find a more specific source. This possibility accounts for certain current insurance industry exclusions for unauthorized or illegal biotechnology-related presence in insurance policies for claims by grain elevators, as the "blending trail" may stop at the elevators.

Several traceability systems are evolving today from both ends of the supply chain. Traceability software being developed and offered ranges in price from $150,000 to $4,000,000, depending on the depth of data needed. Systems that are designed from the farm level capture all details from type of seed, agronomic practices, handling, lists of all parties who dumped into common bin, unit numbers, etc. all the way to door-of-plant or feeding location. Such farm level systems are being marketed on a cost-per-bushel basis or through licensing of software to capture data. These systems thus far are costly, require entry of a great deal of data, and are focused on sectors of supply chain (farmers and elevators) that cannot or will not pay due to small margins. It is generally difficult to get upstream players to pay for these systems as they are not required today and often include more information than is needed. By contrast, the food industry and retailers already have traceability systems. They would probably desire or demand extension of their systems back up the supply chain rather than adopt a new farm-level-based system, as their costs have already been incurred and their systems are already integrated through to the consumer.

D. Contracting and Liability

Contracting requirements and liability impacts vary widely depending on position in the supply chain and the type of product or ingredient involved.

In the seed sector, quality standards are high. Testing is performed on all production. Production and testing records are kept for a minimum of three years as required by the Federal Seed Act.

At the next stage, farmers usually market their production to an elevator network or directly to processors or feeders. A substantial portion of grain is also used directly on the farm for animal feeding. Contracts are governed by industry trade rules that have evolved with industry practices and most often prevail in the case of the commercial markets. Industry arbitration by peers, rather than legal action, is the normal method of dispute resolution. Liability can typically only be passed back if fraud or the occurrence of a specific incident that renders a product non-merchantable is subsequently discovered, since quality is defined by testing at time of title transfer. Discounts and premiums usually apply relative to contracted quality to adjust value to the market. The elevator system always strives to find markets for all grain produced in a particular growing season, regardless of quality.

The introduction of biotech grain and grain products and their potential liabilities may affect the insurance that certain members of the food and feed chain can obtain for business-related risks. Certain insurance companies have recently added exclusions to their policies for unauthorized biotechnology-related presence. With such new policies, there is uncertainty about who in the food and feed chain will bear the liability for a biotechnology related claim, which could have significant impacts on the marketplace.

Farther down the supply/value chain, buyers typically set minimum quality standards accepted or specific traits required or excluded under contracts. Lower quality materials are often rejected based on testing. Acceptance of product in most cases also implies that liability/risk transfer occurs. Tolerances or any other attributes of value such as process verification are usually covered in the contract. Liabilities that arise are typically for non-performance on shipment time or quantity or for non-payment, as testing handles quality issues within contracted tolerances.

Unlike the conditions for all the upstream transactions in the food chain, food manufacturers and retailers do not have contracts with consumers. Liability is defined on the basis of federal regulations/standards on product content, safety, labeling, etc. Consumer expectations as to their right to know certain information about food products purchased can create commercial risks for food manufacturers and retailers. Such risks can arise when consumers expect "zero tolerance" for certain materials, such as transgenic-derived AP, even if those materials have met all applicable government requirements and a "zero tolerance" standard is in practice unattainable.

E. Testing-associated Issues

There are three main categories of tests available for the detection of products of modern biotechnology: 1) direct DNA-based tests to detect the specific DNA sequences (transgenes) inserted into the plant; 2) indirect protein-based tests to detect the specific protein expressed by the new transgene, and 3) bioassays to test for the production of the trait. Each new genetically engineered event contains specific DNA (the transgene) that directs the plant to produce a desired characteristic or trait, plus additional DNA in flanking regions of the transgene that is required for the production of that desired trait, and, most often, additional marker genes used for the initial selection of the transgenic plants in the early stages of product development.

DNA-based tests such as those that use polymerase chain reaction (PCR) amplification include methods that either detect specific DNA sequences from the transgene or associated flanking regions.

Enzyme-linked immunosorbent assays (ELISAs) are antibody-mediated tests that detect the binding of the specific expressed protein product of an inserted gene in sample material to antibodies specific to the expressed protein attached to a fixed surface such as a microtiter dish. Lateral flow strips are a specific adaptation of ELISA technology using specific antibodies immobilized on a small nitrocellulose strip, on which binding of the expressed proteins is coupled to a colour reaction. Both types of tests are relatively fast.

Protein-based tests such as ELISAs and lateral flow strips tend to be simpler and less expensive than PCR but in some cases they can not identify the specific genetic change (the molecular insertional event). Similarly, ELISAs may not be able to distinguish between events producing the same protein. This is a significant consideration since the identification of the specific event by DNA-based testing may be required to support regulations pertaining to regulatory compliance, food labelling or verification of claims pertaining to grain and food products.

Compared to seed and grain, detecting transgenic constituents in multi-ingredient products or foods is complicated by the effects of food processing. Processing can affect the sensitivity and validity of the assays and increases the technical complexity of detecting low levels of transgenic constituents in low percentage ingredients.

Although testing methods are currently available for the detection of specific traits in numerous plants, crops, and some foods, most methods are not internationally accepted and may have not been validated based on international standards. Additionally, no single test can presently detect all events. For routine use in international trade and compliance testing, these methods must yield predictable results. To meet this goal, methods should be validated and refined to address issues associated with method variability and comparability. This includes the development and use of appropriate controls and an accurate determination of the sensitivity and specificity of each method, the development and mutual recognition of reference materials, and harmonization of international standards related to testing.

In addition, the validation process must consider the different types of products being tested (e.g., corn leaves, corn seed, corn pollen, corn flour, corn chips, cornstarch, corn oil and corn sugar). Since products are tested along the supply chain and may undergo processing or compositional changes during the IdP process, it is critical that the grain and grain product test results are comparable for the percent transgenic material present in the test sample, and appropriate sampling protocols are followed.

Setting standard methods for testing samples for the presence of transgenic material in seed, grain and grain products is complicated by a number of factors:

- There is no international consensus on testing methods and thus there are no international standards for testing for transgenic material in products.
- Accurate testing requires defined limits of specific analytical measurements including required sensitivity, sample size, uncertainty measurements, and appropriate validation criteria.
- There are many possible products from transgenic plants, plant parts, and processed foodstuffs, including highly refined substances such as vegetable oil containing little or no detectable transgene-derived protein or DNA.
- Products from transgenic plants and from other transgenic or non-transgenic plants may be combined in varying proportions during processing or handling.
- Inspection, conformity and compliance plans need to be developed on a product-by-product basis for a large number of products.
- Both non-specific, qualitative detection methods (sufficient to address the claim "may contain transgenic or GM products") and specific, quantitative detection methods (for percent determinations) need to be developed to meet different testing needs. Different tests exhibit different practical limitations

and differences in sensitivity and specificity. A number of external factors beyond materials initially planted affect the results that will be obtained upon testing.

In the absence of international standards and harmonization of testing between regulatory authorities, there is significant variability associated with testing grain and food testing for transgenic content. Testing of the same product may produce different results under different test methodologies, at different places in the production chain, or in the final food. As such, there are commercial risks associated with supplying transgenic (or non-transgenic) ingredients based upon testing results.

This testing issue is associated with the lack of "predictability" in the marketplace. The lack of predictability in available testing methods creates commercial risk. This means that a given product may not predictably test "within" the specification of the contract all along the supply chain. The different results may not be associated with commingling of the product with transgenic commodity material, but rather with the statistics of sampling or differences in the sampling method, testing strategies, or capabilities of specific test methods. The possibility exists that a grain product grown from compliant seed, delivered as compliant grain to a processing facility, processed to produce a compliant grain product, and incorporated into the finished food will not test "in compliance" under another compliance testing program. This possibility illustrates the significant commercial risks associated with the supply and marketing of food products in countries that have specific regulations pertaining to foods derived from modern biotechnology, because it may not be possible to effectively manage compliance in those markets.

The development of international standards for test methods and the mutual recognition of reference materials would reduce the variability in testing related to international trade of grain and grain products with respect to transgenic materials. Reducing the variability between and within test methods will increase the predictability of testing compliance in these products and reduce the commercial risks associated with these products. The availability of internationally accepted methods may be facilitated through international organizations and formal and informal cooperation between industry and government organizations within countries.

F. Discrepancy between Regulatory and Commercial Requirements

A country's regulatory framework for mandatory labeling of genetically engineered products most often specifies the requirements for a product to comply with labeling or be exempt from labeling. These requirements, in many cases, serve as the minimum contractual specifications for grain and grain products entering the country for use in food and feed.

In countries with mandatory labeling for transgenic-derived products, there are few food manufacturers who are willing to label their products as genetically engineered or "GM." In most cases, food manufacturers who sell products in countries with mandatory labeling either reformulate and/or substitute with products from non-engineered crops or purchase from supply chains that allow the ingredients to be exempt from mandatory labeling (based on IdP and compliance with a specified AP tolerance). Another potential option for food manufacturers is to relocate manufacturing operations.

In cases where food manufacturers and ingredient

suppliers are sourcing ingredients that are exempt from labeling, the contractual specifications can vary depending on the level of the manufacturers' commercial risk aversion. In some cases, the contractual specifications are identical to the regulatory requirements to be exempt from labeling. If the level of risk aversion is greater, then some food manufacturers require product specifications in excess of the regulatory requirement. For example, in the United Kingdom it is common that the specifications for non-genetically engineered ingredients are IdP and a 0.1 percent AP limit, versus the 0.9 percent AP regulatory requirement for the EU as a whole.

The same variability exists in the requirements for identity preservation systems for exempt products. In general, there are minimum standards for the stringency of identity preservation that must be executed to qualify for product exemption from labeling. In some cases, the expectations are clearly defined through government guidance documents (e.g., Australia), while others may be spelled out by industry groups (e.g., the British Retail Consortium) or business councils. Government standards are readily adopted, while industry group guidance may be regionally adopted or implemented only by specific manufacturers. To reduce the commercial risk, some of these industry standards exceed the legal requirements. We wish to note that it is not just countries that could have a significant input and impact on labeling and the marketplace. Mega-retailers and mega-customers could also drive rapid change in labeling.

G. Impacts of Market Segmentation

Market segmentation is the process of dividing a larger market into several sub-markets. Market segmentation arises primarily out of the processes of capturing value or managing commercial risk. Some basic premises come into play:

- The tighter the product requirements (whether in terms of tolerances, regulatory requirements, demonstration of end use benefits), the more "value" segregates.
- The more "value" segregates, the more IdP must increase to insure the market captures the value created.
- Often, as the value of a product increases, so does the commercial risk, leading to the desire for more testing and more control.
- More control implies more tracing. This increases the probability or at least opens the door for liability transfer further back in the supply chain if an "incident" destroys value capture for succeeding parties.
- Increased commercial risk or control drives a shift toward integration or a dramatic narrowing of the group of suppliers. For some downstream entities, "bigger is usually better" in terms of their suppliers as it is easier to deal with a limited number or a single supplier than a broad market.
- The value of products becomes less transparent to others in the market.
- The market power of individual players dictates market rules and the sharing of value. Substitutability of products dilutes market power.

All of the above create differential returns to growers and different levels of access to particular markets by growers of different specialty varieties of the same crop.

Markets can segment for reasons that are not easily measurable. For transgenic products, market segmentation is further complicated by differing attitudes among the population of product developers, producers, suppliers, and consumers that range from the view that current products are very similar to traditional products and future products offer the promise of addressing future needs in the U.S. and elsewhere to a core concern about inherent differences between biotechnology-derived and traditional products and a focus on possible longer-term safety concerns or consequences to the environment.

Various sectors of the market view the segmentation issue or are affected by it in different ways:

Commodity system – This represents the largest sector of the market, one that relies on harmonization of rules and regulations to allow commerce to flow efficiently. Liability transfer has traditionally been defined through negotiated contracts or industry trade rules. Product commingling is unavoidable given high volumes, high speed, and the need to blend to meet defined quality standards. Margins tend to be small given large number of market participants, homogeneity of product, free-flowing information, and limited or no barriers to entry. Buyers often gravitate to broad, common quality specifications, as price spreads are often not wide enough to support several differentiated products. Any product segmentation is usually driven by broad quality factors such as oil content, protein level, etc.

Political system – Different countries base their traceability and labeling policies and regulations on different approaches, which may be more or less science-based or market–based and may blend in various "right to know" concepts or include various measures for domestic economy protection. Issues of harmonization, liability transfer, etc. evolve very slowly through trade agreements, WTO rulings, and/or compromises through international groups. Multiple other external factors such as wars, political differences, economic imbalances, etc. also affect progress toward harmonization. Markets can segment via political or economic coalitions.

Food system – Before products are sold in any country, food manufacturers need clear, understandable guidelines as to what is necessary to meet all legal and regulatory requirements. However, in addition, consumer desires will drive actions of food manufacturers. Most food manufacturers believe strongly in a demand-pull marketplace, although current products of modern biotechnology in agriculture have not entered the marketplace via demand-pull. Lack of harmonization globally is forcing diversification of the food product manufacturing base to fit differing market needs. Trends are more towards "right to know" on a number of parameters, especially on nutrient content and origin. Growth in markets for differentiated products will intensify price segregation to participants in the supply chain. These price signals both communicate market needs and spread the distribution of value, driving more segmentation, more integration, more contracting, tighter tracing/ testing, and a general narrowing of the supply chain to insure the delivery of (specified) value to the end consumer.

H. Consumer Implications of Various Labeling Regimes

Consumers need access to certain basic information to make informed choices about food products they purchase, including nutritional composition and price. Consumer purchasing decisions are also influenced by personal and cultural factors such as familiarity, taste, and religious preferences. As markets have become more global, some consumers have sought additional information, such as where a product was produced, what environmental and employment practices were followed, and what processing methods were used. The food and feed supply chain has sought to establish economically efficient identity preservation and other marketing systems to provide the desired information.

Some countries have established mandatory labeling requirements for most food products that have been produced using biotechnology. These regulatory requirements and commercial decisions affect the amount and type of information available to retail consumers, and may influence product availability and choice. Any resulting cost increase is absorbed throughout the food system, from producer through final consumer.

Mandatory labeling for genetically engineered products has effectively prevented most biotech products from being marketed in countries with such regulations. In the U.S., where there is no mandatory process-based food labeling, many foods contain genetically engineered ingredients and no health or safety issues have arisen. Consumers purchase these products, frequently unaware that they were made using biotech ingredients. U.S. consumers who want non-biotech products have reduced choices that are generally higher priced.

Consumer research shows a substantial majority of U.S. consumers want more information about their food. When U.S. consumers are asked if they want food containing genetically engineered products labeled, they usually say "yes." If they are not asked specifically, biotechnology labeling is rarely mentioned. The food label provides a simple way for consumers to get information although it may also be acquired at the point of retail purchase, by contacting the manufacturer, and increasingly, by seeking information on the internet.

IV. POLICY CONCERNS/ISSUES RAISED BY THE ANALYSIS OF THE CURRENT SITUATION

These are some of the key policy concerns and issues that emanate from the Committee's discussions and deliberations captured in the full report.

- The complexity of complying with multiple labeling and traceability requirements that differ by market and country imposes additional costs and inefficiencies on the supply chain.
- Some tracing and labeling requirements are uncertain, the systems in some countries are rapidly evolving, and some systems lack specifics about what constitutes compliance. This increases commercial challenges for the production, marketing, and trade of grain and grain products.
- There are inherent difficulties in complying with biotechnology-specific traceability and labeling regulations governing product characteristics that cannot be detected. Government-imposed process certification requirements pose challenges for industry.
- There are significant sampling and testing issues when one is supplying grain or grain products that are required to meet specific tolerances or thresholds. This difficulty reflects the sampling and analytical challenges associated with the testing of grain for genetically-engineered content. Variability between tests relates to the absence of reference standards and internationally-validated and harmonized test methods. Without harmonization and standardization it will remain difficult to practically and reproducibly determine whether products are

in compliance with a specific tolerance. The problem of variability among tests is exacerbated as tolerance levels decrease.

- Which test results take precedence in the movement of grain or grain products needs to be clarified. Mutual recognition of testing and sampling methods may enable downstream parties to accept testing at the origin of a grain or grain product without the need for additional testing. Development and recognition of testing certificates would reduce the commercial risks along the supply chain.
- Industry is grappling with how to address issues related to "acceptable zero" versus an absolute zero for both approved and unapproved events.
- The commercial marketplace is struggling with varying concepts of what product specifications (tolerances) trigger labeling of genetically engineered versus non-engineered products. Even where specific tolerances are provided by regulation, the marketplace may require suppliers to provide products at much lower, sometimes zero, tolerance levels, to allow them to make a zero claim to the eventual buyers, particularly on consumer products.
- Uncertainties as to liability transfer and exclusions for biotechnology-related claims by some insurance companies for certain portions of the supply chain can have significant impacts on the marketplace.
- There is a role for international organizations, as well as bilateral and multilateral efforts, to address some of these impacts and implications, but it not always clear which approach is best suited for the resolution of each. The U.S. government should continue to creatively explore different venues and approaches.

Consumers' interest in seeking additional information about their food is increased by variability in retail requirements in different countries and associated debates regarding safety, environmental concerns, and general acceptability of biotech crops.

The Chips are Coming

Paranoid ranting is a staple on the internet, and the topic of biochips is a particular favorite. Biochips—any microprocessor chips that can be used in biology—mean many things to many people. The web's preachers focus on a simple idea: an implanted chip that identifies each person uniquely, and can be used to track their location.

This wonderful idea—capable of locating lost children, downed soldiers, and wandering Alzheimers patients—sounds alarming even to many mild-mannered civil libertarians. To Living Truth Ministries of Austin, TX, it is no less than "a tremendous bonanza for the coming Antichrist!" Second Advent Ministries of San Bernardino, Calif., warns us that the US social security number will soon be replaced (at the orders of the "New World Order United Nations-controlled world government", no less) by a chip-encoded "18-digit numbering system [which] will consist of THREE separate sets of SIX DIGITS each: 6-6-6! Coincidental? No. Not if you believe Bible prophecy."

Reality is somewhat less alarming. A simple ID chip is already walking around in tens of thousands of individuals, but all of them are pets. Companies such as AVID (Norco, Calif.), Electronic ID, Inc. (Cleburne, Tx.), and Electronic Identification Devices, Ltd. (Santa Barbara, Calif.) sell both the chips and the detectors. The chips are the size of an uncooked grain of rice, small enough to be injected under the skin using a syringe needle. They respond to a signal from the detector, held just a few feet away, by transmitting out an identification number. This number is then compared to database listings of registered pets.

Daniel Man, a plastic surgeon in private practice in Boca Raton, Fla., holds a patent on a more sinister device: a chip that would enable lost humans to be tracked by satellite. He has a large-scale model, but for miniaturization and regulatory approval he will need a substantial amount of cash. "There are some companies that are interested, but nothing is official," he says. "We're not giving up on it yet."

Chips to Monitor Your Every Footstep and Body Fluid

The civil liberties debate over biochips has obscured their more ethically benign and medically useful applications. Medical researchers have been working to integrate chips and people for many years, often plucking devices from well-known electronic appliances. Jeffrey Hausdorff of the Beth Israel Deaconess Medical Center (Boston, Mass.) has used the type of pressure-sensitive resistors found in the buttons of a microwave oven as stride timers. He places one sensor in the heel of a shoe, and one in the toe, and adds a computer to the ankle to calculate the duration of each stride.

"Young, healthy subjects can regulate the duration of each step very accurately," he says. But elderly patients prone to frequent falls have extremely variable stride times, a flag that could indicate the need for more strengthening exercises or a change in medication. Hausdorff is also using the system to determine the success of a treatment for congestive heart failure. By monitoring the numbers of strides that a person takes, he can directly measure the patient's activity level, bypassing the often flawed estimate made by the patient.

Hausdorff's chips are external, but a chip proposed by

Sensors for Medicine and Science, Inc. (S4MS; Germantown, Md.) will be injected under the skin. The chip will allow diabetics to easily monitor the level of the sugar glucose in their blood. Diabetics currently use a skin prick and a hand-held blood test, and then medicate themselves with insulin depending on the result. The system is simple and works well, but the need to draw blood means that most diabetics don't test themselves as often as they should. Although they may get away with this in the short term, in later life those who monitored infrequently suffer from blindness, loss of circulation, and other complications.

The solution is more frequent testing, using a less invasive method. Various companies are trying to extract the required body fluid, using either an electric current (Cygnus, Inc., Redwood City, Calif.) or a tiny pore induced by a laser (SpectRx, Norcross, Ga.) or a small needle (Integ, St Paul, Minn.).

"All of these approaches ameliorate the drawing of blood," says George Daniloff, Chief Scientific Officer of S4MS, "but they still require manipulation of skin, and the skin may be perturbed." The S4MS chip, however, will sit underneath the skin, sense the glucose level, and send the result back out by radio-frequency communication.

A light-emitting diode (LED) starts off the detection process. The light that it produces hits a fluorescent chemical: one that absorbs incoming light and re-emits it at a longer wavelength. The longer wavelength of light is then detected, and the result is sent to a control panel outside the body.

Glucose is detected because the sugar reduces the amount of light that the fluorescent chemical re-emits. The more glucose there is, the less light that is detected. S4MS is still developing the perfect fluorescent chemical, but the key design innovation of the S4MS chip has been fully worked out. The idea is simple: the LED is sitting in a sea of the fluorescent molecules. In most detectors the light source is far away from the fluorescent molecules, and the inefficiencies that come with that mean more power and larger devices. The prototype S4MS chip uses a 22μW LED, almost forty times less powerful than the tiny power-on buttons on a computer keyboard. The low power requirements mean that energy can be supplied from the outside, by a process called induction. The fluorescent detection itself does not consume any chemicals or proteins, so the device is self sustaining.

Daniloff has a working model of an oxygen sensor that. uses the same layout. With it's current circuitry it is about the size of a large shirt-button, but the final silicon wafer will be less than a millimeter square. "The scale transition is not a challenge," says Daniloff. "The challenge is to illustrate the principle." The oxygen sensor will be useful not only to monitor breathing in intensive care units, but also to check that packages of food, or containers of semiconductors stored under nitrogen gas, remain airtight.

Tom Ferrell of the Oak Ridge National Laboratory in Oak Ridge, Tenn., is also developing an oxygen-sensing chip, but his version sends light pulses out into the body. The light is absorbed to varying extents, depending on how much oxygen is being carried in the blood, and Ferrell's chip detects the light that is left. The rushes of blood pumped by the heart are also detected, so the same chip is a pulse monitor. A number of companies already make large-scale versions of such detectors, and Ferrell says he has "collaborated with almost every company in this area," although he will not name names.

Ferrell estimates that the oxygen chip is two years away, but he has already reduced the dimensions of his temperature-sensing chip to 3mm per side. The transition of certain semiconductors to their conducting state is inherently sensitive to temperature, so designing the sensor was simple enough. With some miniature radio-frequency transmitters, and foam-rubber earplugs to hold the chip in place, the device is complete. Ferrell sees applications for anyone from sick children, to chemotherapy patients who can be plagued by sudden rises in body temperature in response to their anti-cancer drugs.

Brain Surgery with an On-off Switch

Sensing and measuring is one thing, but can we switch the body on and off? Heart pacemakers use the crude approach: large jolts of electricity to synchronize the pumping of the heart. The electric pulses of the Activa implant, made by Medtronic, Inc. (Minneapolis, Minn.), are directed not at the heart but the brain. They turn off brain signals that cause the uncontrolled movements, or tremors, associated with diseases such as Parkinson's. Drug therapy for Parkinson's disease aims to replace the brain messenger dopamine, a product of the brain cells that are dying. But eventually the drug's effects wear off, and the erratic movements come charging back. "I've had patients tell me that they scare their grandkids because they are flaying around like a fish," says Gary Heit, a neurosurgeon at Stanford University Medical Center. "Imagine your whole body twisted like a charleyhorse."

The next option is to remove a portion of the thalamus, a brain structure that triggers the tremors. The implant, cleared for use in the United States in August 1997, is a new alternative that uses high-frequency electrical pulses to reversibly shut off the thalamus. "It has the same benefits as thalamotomy [the brain operation] without the risks," says Heit. The implantation surgery is far less traumatic than thalamotomy, and if there are any post-operative problems the stimulator can simply be turned off.

Adding Back the Senses

The Activa implant interferes with aberrant brain functioning. The most ambitious bioengineers are trying to add back brain functions, restoring sight and sound where there was darkness and silence.

The success story in this field is the cochlear implant. Most hearing aids are glorified amplifiers, but the cochlear implant is for patients who have lost the hair cells that detect sound waves. For these individuals, no amount of amplification is enough.

The cochlear implant delivers electrical pulses directly to the nerve cells in the cochlea, the spiral-shaped structure that translates sound into nerve pulses. In normal hearing individuals, sound waves set up vibrations in the walls of the cochlea, and hair cells detect these vibrations. High frequency noises (deep notes) vibrate the base of the cochlea, while low frequency notes vibrate nearer the top of the spiral. The implant mimics the job of the hair cells. It splits the frequencies of incoming noises into a number of channels (typically eight), and then stimulates the appropriate part of the cochlea.

The two most successful implants are the Clarion (developed at the University of California at San Francisco (UCSF), Research Triangle Institute, N.C., and Advanced Bionics Corporation of Sylmar, Calif.) and the Nucleus (developed at the University of Melbourne, Australia, and made by Cochlear of Sydney, Australia). Upgrades largely focus on improving the speech processing software, which is operated by a minicomputer worn on the patient's belt. Theoretically, increasing the number of channels (and electrodes) could improve sound perception. But speech is perceived in an area of the cochlea only 14 mm long, and spacing the electrodes too close to each other causes signals to bleed from one channel to another.

The result is a broad-brush version of hearing. While some recipients of the devices report speech-like sounds, many characterize their new world as being populated with legions of quacking ducks or banging garbage cans. But the success is undeniable. "Currently two thirds to three quarters of patients [with more recent models] can understand speech without lip-reading," says Steve Rebscher, a member of the UCSF team. "It's pretty amazing, and certainly better than a lot of people anticipated these devices would do."

With the ear at least partially conquered, the next logical target is the eye. Several groups are working on implantable chips that mimic the action of photoreceptors, the light-sensing cells at the back of the eye. Photoreceptors are lost in both retinitis pigmentosa, a genetic disease, and age-related macular degeneration, the most common cause of lost sight in the developed world.

Joseph Rizzo of the Massachusetts Eye and Ear Infirmary, and John Wyatt of the Massachusetts Institute of Technology have made a twenty-electrode, 1mm-square chip, and implanted it at the back of rabbits' eyes. The original chip, the thickness of a human hair, put too much stress on the eye, so the new version is ten times thinner. The final set-up will include a fancy camera mounted on a pair of glasses. The camera will detect and encode the scene, then send it into the eye as a laser pulse, with the laser also providing the energy to drive the chip. Rizzo has confirmed that his tiny array of light receivers (photodiodes) can generate enough electricity to run the chip. He has also found that the amount of electricity needed to fire a nerve cell into action is ~100-fold lower in the eye than in the ear, so the currents can be smaller, and the electrodes more closely spaced.

For now the power supply comes from a wire inserted directly into the eye and, using this device, Rizzo has detected signals reaching the brain. "We know we can get visual information to the brain," he says, "but we don't know what kind of vision that will result in."

Eugene de Juan of Johns Hopkins Wilmer Eye Institute (Baltimore, Md.) is trying to answer that question by using human subjects. His electrodes, inserted directly into the eye, are large and somewhat crude. But his results have been startling. Completely blind patients have seen well-defined flashes, which change in position and brightness as de Juan changes the position of the electrode or the amount of current. In his most recent experiments, patients have identified simple shapes outlined by multiple electrodes. With as little as an 8 x8 array, de Juan believes he could approximate character recognition, and a 25 x 25 array might give a crude image.

The big money in eye implants is in Germany, where the government has pledged ~$10 million to two projects. One is similar to the US projects, in which chips are implanted on the surface of the retina, the structure at the back of the eye. The other project is putting its implants at the back of the retina, where the photoreceptors are normally found. These 'subretinal' chips may block the transport of oxygen and food to the overlying nerve cells, so Eberhart Zrenner of the University of Tübingen, Germany, is developing 'chain mail' electrode arrays, with plenty of holes for the delivery of supplies.

The subretinal approach more closely mimics the action of the lost photoreceptors. The chip delivers its message at the source, before intervening layers of retinal nerve cells do some complicated processing. Rizzo thinks that his chip's hardware will be able to compensate for these computations, but for now he has other things to worry about. "There are all sorts of processes that go on that we're just going to ignore initially," he says. As with the cochlear implant, he says, "the biological tests will show you the best way to stimulate. Ultimately the only thing that matters is what the person sees."

The first big test—showing that a remotely powered device can give true visual signals—is yet to be done. "We haven't got to first base yet," says Rizzo, "but we're close." With a working device many years away, Rizzo is a model of caution. "It's always better to talk down the possibilities than to talk them up, because there are many reasons why the implants might fail," he says. "This is science fiction stuff."

Selling DNA Data

Human cells are almost ridiculously tiny and efficient. Every one of them has an entire genetic instruction set—three billion As, Cs, Gs and Ts—packed into a nucleus one hundredth of a millimeter across. And a few cents worth of carbon, nitrogen and phosphorus will keep millions of cells happy in a dish, reproducing themselves and their DNA data.

Incyte Pharmaceuticals, Inc. (Palo Alto, Calif.) also stores DNA data, but in a way that is somewhat less compact and far more likely to impress a computer geek. Ground-zero for Incyte is a highly air-conditioned room filled with scores of black, towering supercomputers worth up to a million dollars each.

Nine of the ten largest pharmaceutical companies in the world each pay Incyte ~$5 million per year to look at the data on those computers. The money has brought an expanding workforce—from 160 to 675 employees in the last two and a half years—and a frantic effort to keep generating more and more information. "We're maybe a year away from going bankrupt if we don't pursue new technologies at any one point," says Tod Klingler, Director of Research Bioinformatics at Incyte.

The manpower and computer power is needed to tame an ever-growing mountain of DNA sequence information. "We now have 24 organisms completely sequenced we have their entire genetic message written down," says Temple Smith of Boston University. "This information allows us to get information about one organism and use that to understand another."

The Chips are Coming

Paranoid ranting is a staple on the internet, and the topic of biochips is a particular favorite. Biochips—any microprocessor chips that can be used in biology—mean many things to many people. The web's preachers focus on a simple idea: an implanted chip that identifies each person uniquely, and can be used to track their location.

This wonderful idea—capable of locating lost children, downed soldiers, and wandering Alzheimers patients—sounds alarming even to many mild-mannered civil libertarians. To Living Truth Ministries of Austin, TX, it is no less than "a tremendous bonanza for the coming Antichrist!" Second Advent Ministries of San Bernardino, Calif., warns us that the US social security number will soon be replaced (at the orders of the "New World Order United Nations-controlled world government", no less) by a chip-encoded "18-digit numbering system [which] will consist of THREE separate sets of SIX DIGITS each: 6-6-6! Coincidental? No. Not if you believe Bible prophecy."

Reality is somewhat less alarming. A simple ID chip is

already walking around in tens of thousands of individuals, but all of them are pets. Companies such as AVID (Norco, Calif.), Electronic ID, Inc. (Cleburne, Tx.), and Electronic Identification Devices, Ltd. (Santa Barbara, Calif.) sell both the chips and the detectors. The chips are the size of an uncooked grain of rice, small enough to be injected under the skin using a syringe needle. They respond to a signal from the detector, held just a few feet away, by transmitting out an identification number. This number is then compared to database listings of registered pets.

Daniel Man, a plastic surgeon in private practice in Boca Raton, Fla., holds a patent on a more sinister device: a chip that would enable lost humans to be tracked by satellite. He has a large-scale model, but for miniaturization and regulatory approval he will need a substantial amount of cash. "There are some companies that are interested, but nothing is official," he says. "We're not giving up on it yet."

Chips to Monitor Your Every Footstep and Body Fluid

The civil liberties debate over biochips has obscured their more ethically benign and medically useful applications. Medical researchers have been working to integrate chips and people for many years, often plucking devices from well-known electronic appliances. Jeffrey Hausdorff of the Beth Israel Deaconess Medical Center (Boston, Mass.) has used the type of pressure-sensitive resistors found in the buttons of a microwave oven as stride timers. He places one sensor in the heel of a shoe, and one in the toe, and adds a computer to the ankle to calculate the duration of each stride.

"Young, healthy subjects can regulate the duration of each step very accurately," he says. But elderly patients prone to frequent falls have extremely variable stride times, a flag that could indicate the need for more strengthening exercises or a change in medication. Hausdorff is also using the system to determine the success of a treatment for congestive heart failure. By monitoring the numbers of strides that a person takes, he can directly measure the patient's activity level, bypassing the often flawed estimate made by the patient.

Hausdorff's chips are external, but a chip proposed by Sensors for Medicine and Science, Inc. (S4MS; Germantown, Md.) will be injected under the skin. The chip will allow diabetics to easily monitor the level of the sugar glucose in their blood. Diabetics currently use a skin prick and a hand-held blood test, and then medicate themselves with insulin depending on the result. The system is simple and works well, but the need to draw blood means that most diabetics don't test themselves as often as they should. Although they may get away with this in the short term, in later life those who monitored infrequently suffer from blindness, loss of circulation, and other complications.

The solution is more frequent testing, using a less invasive method. Various companies are trying to extract the required body fluid, using either an electric current (Cygnus, Inc., Redwood City, Calif.) or a tiny pore induced by a laser (SpectRx, Norcross, Ga.) or a small needle (Integ, St Paul, Minn.).

"All of these approaches ameliorate the drawing of blood," says George Daniloff, Chief Scientific Officer of S4MS, "but they still require manipulation of skin, and the skin may be perturbed." The S4MS chip, however, will sit underneath the skin, sense the glucose level, and send the result back out by radio-frequency communication.

A light-emitting diode (LED) starts off the detection process. The light that it produces hits a fluorescent chemical: one that absorbs incoming light and re-emits it at a longer wavelength. The longer wavelength of light is then detected, and the result is sent to a control panel outside the body.

Glucose is detected because the sugar reduces the amount of light that the fluorescent chemical re-emits. The more glucose there is, the less light that is detected. S4MS is still developing the perfect fluorescent chemical, but the key design innovation of the S4MS chip has been fully worked out. The idea is simple: the LED is sitting in a sea of the fluorescent molecules. In most detectors the light source is far away from the fluorescent molecules, and the inefficiencies that come with that mean more power and larger devices. The prototype S4MS chip uses a 22μW LED, almost forty times less powerful than the tiny power-on buttons on a computer keyboard. The low power requirements mean that energy can be supplied from the outside, by a process called induction. The fluorescent detection itself does not consume any chemicals or proteins, so the device is self sustaining.

Daniloff has a working model of an oxygen sensor that uses the same layout. With it's current circuitry it is about the size of a large shirt-button, but the final silicon wafer will be less than a millimeter square. "The scale transition is not a challenge," says Daniloff. "The challenge is to illustrate the principle." The oxygen sensor will be useful not only to monitor breathing in intensive care units, but also to check that packages of food, or containers of semiconductors stored under nitrogen gas, remain airtight.

Tom Ferrell of the Oak Ridge National Laboratory in Oak Ridge, Tenn., is also developing an oxygen-sensing chip, but his version sends light pulses out into the body. The light is absorbed to varying extents, depending on how much oxygen is being carried in the blood, and Ferrell's chip detects the light that is left. The rushes of blood pumped by the heart are also detected, so the same chip is a pulse monitor. A number of companies already make large-scale versions of such detectors, and Ferrell says he has "collaborated with almost every company in this area," although he will not name names.

Ferrell estimates that the oxygen chip is two years away, but he has already reduced the dimensions of his temperature-sensing chip to 3mm per side. The transition of certain semiconductors to their conducting state is inherently sensitive to temperature, so designing the sensor was simple enough. With some miniature radio-frequency transmitters, and foam-rubber earplugs to hold the chip in place, the device is complete. Ferrell sees applications for anyone from sick children, to chemotherapy patients who can be plagued by sudden rises in body temperature in response to their anti-cancer drugs.

Brain Surgery with an On-off Switch

Sensing and measuring is one thing, but can we switch the body on and off? Heart pacemakers use the crude approach: large jolts of electricity to synchronize the pumping of the heart. The electric pulses of the Activa implant, made by Medtronic, Inc. (Minneapolis, Minn.), are directed not at the heart but the brain. They turn off brain signals that cause the uncontrolled movements, or tremors, associated with diseases such as Parkinson's.

Drug therapy for Parkinson's disease aims to replace the brain messenger dopamine, a product of the brain cells that are dying. But eventually the drug's effects wear off, and the erratic movements come charging back. "I've had patients tell me that they scare their grandkids because they are flaying around like a fish," says Gary Heit, a neurosurgeon at Stanford University Medical Center. "Imagine your whole body twisted like a charleyhorse."

The next option is to remove a portion of the thalamus, a brain structure that triggers the tremors. The implant, cleared for use in the United States in August 1997, is a new alternative that uses high-frequency electrical pulses to reversibly shut off the thalamus. "It has the same benefits as thalamotomy [the brain operation] without the risks," says Heit. The implantation surgery is far less traumatic than thalamotomy, and if there are any post-operative problems the stimulator can simply be turned off.

Adding Back the Senses

The Activa implant interferes with aberrant brain functioning. The most ambitious bioengineers are trying to add back brain functions, restoring sight and sound where there was darkness and silence.

The success story in this field is the cochlear implant. Most hearing aids are glorified amplifiers, but the cochlear implant is for patients who have lost the hair cells that detect sound waves. For these individuals, no amount of amplification is enough.

The cochlear implant delivers electrical pulses directly to the nerve cells in the cochlea, the spiral-shaped structure that translates sound into nerve pulses. In normal hearing individuals, sound waves set up vibrations in the walls of the cochlea, and hair cells detect these vibrations. High frequency noises (deep notes) vibrate the base of the cochlea, while low frequency notes vibrate nearer the top of the spiral. The implant mimics the job of the hair cells. It splits the frequencies of incoming noises into a number of channels (typically eight), and then stimulates the appropriate part of the cochlea.

The two most successful implants are the Clarion (developed at the University of California at San Francisco

(UCSF), Research Triangle Institute, N.C., and Advanced Bionics Corporation of Sylmar, Calif.) and the Nucleus (developed at the University of Melbourne, Australia, and made by Cochlear of Sydney, Australia). Upgrades largely focus on improving the speech processing software, which is operated by a minicomputer worn on the patient's belt. Theoretically, increasing the number of channels (and electrodes) could improve sound perception. But speech is perceived in an area of the cochlea only 14 mm long, and spacing the electrodes too close to each other causes signals to bleed from one channel to another.

The result is a broad-brush version of hearing. While some recipients of the devices report speech-like sounds, many characterize their new world as being populated with legions of quacking ducks or banging garbage cans. But the success is undeniable. "Currently two thirds to three quarters of patients [with more recent models] can understand speech without lip-reading," says Steve Rebscher, a member of the UCSF team. "It's pretty amazing, and certainly better than a lot of people anticipated these devices would do."

With the ear at least partially conquered, the next logical target is the eye. Several groups are working on implantable chips that mimic the action of photoreceptors, the light-sensing cells at the back of the eye. Photoreceptors are lost in both retinitis pigmentosa, a genetic disease, and age-related macular degeneration, the most common cause of lost sight in the developed world.

Joseph Rizzo of the Massachusetts Eye and Ear Infirmary, and John Wyatt of the Massachusetts Institute of Technology have made a twenty-electrode, 1mm-square chip, and implanted it at the back of rabbits' eyes. The original chip, the thickness of a human hair, put too much stress on the eye, so the new version is ten times thinner.

The final set-up will include a fancy camera mounted on a pair of glasses. The camera will detect and encode the scene, then send it into the eye as a laser pulse, with the laser also providing the energy to drive the chip. Rizzo has confirmed that his tiny array of light receivers (photodiodes) can generate enough electricity to run the chip. He has also found that the amount of electricity needed to fire a nerve cell into action is ~100-fold lower in the eye than in the ear, so the currents can be smaller, and the electrodes more closely spaced.

For now the power supply comes from a wire inserted directly into the eye and, using this device, Rizzo has detected signals reaching the brain. "We know we can get visual information to the brain," he says, "but we don't know what kind of vision that will result in."

Eugene de Juan of Johns Hopkins Wilmer Eye Institute (Baltimore, Md.) is trying to answer that question by using human subjects. His electrodes, inserted directly into the eye, are large and somewhat crude. But his results have been startling. Completely blind patients have seen well-defined flashes, which change in position and brightness as de Juan changes the position of the electrode or the amount of current. In his most recent experiments, patients have identified simple shapes outlined by multiple electrodes. With as little as an 8 x8 array, de Juan believes he could approximate character recognition, and a 25 x 25 array might give a crude image.

The big money in eye implants is in Germany, where the government has pledged ~$10 million to two projects. One is similar to the US projects, in which chips are implanted on the surface of the retina, the structure at the back of the eye. The other project is putting its implants at the back of the retina, where the photoreceptors are normally found.

These 'subretinal' chips may block the transport of oxygen and food to the overlying nerve cells, so Eberhart Zrenner of the University of Tübingen, Germany, is developing 'chain mail' electrode arrays, with plenty of holes for the delivery of supplies.

The subretinal approach more closely mimics the action of the lost photoreceptors. The chip delivers its message at the source, before intervening layers of retinal nerve cells do some complicated processing. Rizzo thinks that his chip's hardware will be able to compensate for these computations, but for now he has other things to worry about. "There are all sorts of processes that go on that we're just going to ignore initially," he says. As with the cochlear implant, he says, "the biological tests will show you the best way to stimulate. Ultimately the only thing that matters is what the person sees."

The first big test—showing that a remotely powered device can give true visual signals—is yet to be done. "We haven't got to first base yet," says Rizzo, "but we're close." With a working device many years away, Rizzo is a model of caution. "It's always better to talk down the possibilities than to talk them up, because there are many reasons why the implants might fail," he says. "This is science fiction stuff."

Documents of the Tenth Regular Session of the CGRFA
Rome, 8—12 November 2004

CGRFA-10/04/REPORT Report of the Tenth Regular Session of the Commission on Genetic Resources for Food and Agriculture

Working Documents
CGRFA-10/04/1

Draft Provisional Agenda

CGRFA-10/04/2

Provisional Annotated Agenda and Time-table

CGRFA-10/04/3

Overview of the FAO *Global System for the Conservation and Sustainable Utilization of Plant Genetic Resources for Food and Agriculture*, and its potential contribution to the implementation of the International Treaty on Plant Genetic Resources for Food and Agriculture

CGRFA-10/04/4

Report of the Second Session of the Intergovernmental Technical Working Group on Plant Genetic Resources for Food and Agriculture

CGRFA-10/04/5

Follow-up to the recommendations of the Commission on Genetic Resources for Food and Agriculture and its Working Group, regarding Plant Genetic Resources for Food and Agriculture

CGRFA-10/04/5 Add.1

Progress in the preparation of the Second *Report on the State of the World's Plant Genetic Resources for Food and Agriculture*

CGRFA-10/04/6

Report on the International Network of *Ex Situ* Collections under the Auspices of FAO

CGRFA-10/04/7

Progress in the Development of the Global Strategy for the Management of Farm Animal Genetic Resources

CGRFA-10/04/8

Report of the Third Session of the Intergovernmental Technical Working Group on Animal Genetic Resources for Food and Agriculture

CGRFA-10/04/9

Progress Report on the process of the preparation of the *First Report on the State of the World's Animal Genetic Resources* and the *Report on Strategic Priorities for Action*

CGRFA-10/04/9 Add.1

Initiation of the Follow-up Mechanism for the implementation of Strategic Priorities for Action—Animal Genetic Resources

CGRFA-10/04/9 Add.2

Draft Outline for the *First Report on the State of the World's Animal Genetic Resources* including the *Report on Strategic Priorities for Action*

CGRFA-10/04/9 Add.3

Budgetary and other aspects of a possible International Technical Conference on animal genetic resources

CGRFA-10/04/10.1

Report from FAO on its policies, programmes and activities on agricultural biological diversity:(1) Sectorial matters

CGRFA-10/04/10.2

Report from FAO on its policies, programmes and activities on agricultural biological diversity:(2) Cross-sectorial matters

CGRFA-10/04/10.3

Report from FAO on its policies, programmes and activities on agricultural biological diversity:(3) Priority Areas for Inter-disciplinary Action (PAIAs)

CGRFA-10/04/11.1

Reports from international organizations on their policies, programmes and activities on agricultural biological diversity:(1) United Nations and other Inter-governmental Organizations

CGRFA-10/04/11.2

Reports from international organizations on their policies, programmes and activities on agricultural biological diversity:(2) International Agricultural Research Centres of the Consultative Group on International Agricultural Research (CGIAR)

CGRFA-10/04/11.3

Reports from international organizations on their policies, programmes and activities on agricultural biological diversity: (3) International Non-governmental Organizations

CGRFA-10/04/11 Add.1 Rev.1

Reports from international organizations on their policies, programmes and activities on agricultural biological diversity:Reports arrived late for translation

CGRFA-10/04/12

Cooperation with the Convention on Biological Diversity

CGRFA-10/04/13

Progress in the Draft Code of Conduct on Biotechnology as it relates to genetic resources for food and agriculture: policy issues, gaps and duplications

CGRFA-10/04/14

Future work of the Commission

Information Documents

CGRFA-10/04/Inf.1

Statutes of the Commission on Genetic Resources for Food and Agriculture

CGRFA-10/04/Inf.2

Statutes of the Intergovernmental Technical Working

Group on Animal Genetic Resources for Food and Agriculture, and Members elected by the Ninth Regular Session of the Commission

CGRFA-10/04/Inf.3

Statutes of the Intergovernmental Technical Working Group on Plant Genetic Resources for Food and Agriculture, and Members elected by the Ninth Regular Session of the Commission

CGRFA-10/04/Inf.4

Report of the Meeting on the Evaluation of the Pilot Phase of the New Approach for Monitoring the Implementation of the Global Plan of Action for the Conservation and Sustainable Utilization of Plant Genetic Resources for Food and Agriculture

CGRFA-10/04/Inf.5

Indicators and reporting format for monitoring the implementation of the Global Plan of Action for the Conservation and Sustainable Utilization of Plant Genetic Resources for Food and Agriculture

CGRFA-10/04/Inf.6

Country Progress Report on the implementation of the *Global Plan of Action for the Conservation and Sustainable Utilization of Plant Genetic Resources for Food and Agriculture*

CGRFA-10/04/Inf.7

Strengthening Seed Systems

CGRFA-10/04/Inf.8

Country Guidelines for the Second *State of the World's Plant Genetic Resources for Food and Agriculture*

CGRFA-10/04/Inf.9

Draft Report on the Strategic Priorities for Action

CGRFA-10/04/Inf.10

Statement of competence and voting rights submitted by the European Community and its Member States only

CGRFA-10/04/Inf.11

Agreement Between the Tropical Agricultural Research and Higher Education Center (CATIE) and the Food and Agriculture Organization of the United nations (FAO) Placing Collections of Plant Germplasm under the Auspices of FAO

CGRFA-10/04/Inf.13

Reports from International Organizations on their Policies, Programmes and Activities on Agricultural Biological Diversity Part II: The International Agricultural Research Centres of the Consultative Group on International Agricultural Research (CGIAR)—List of National and International Partners in Research Programmes on Farm Animal Genetic Resources

CGRFA-10/04/Inf.14

Report on the International Network of Ex Situ Collections under the Auspices of FAO: Further Information Provided by the International Centre for Tropical Agriculture (CIAT), Regarding its Request for a Re-examination of U.S. Patent No.5,894,079

CGRFA-10/04/Inf.16

Strengthening National Structures for the Management of Farm Animal Genetic Resources—Results of a Questionnaire Survey

Bibliography

Aparicio, S. *et al.*Whole-genome shotgun assembly and analysis of the genome of *Fugu rubripes*. *Science* 297, 1301–1310 (2002).

Avery, O. T., MacLeod, C.M.& McCarty, M. Studies of the chemical nature of the substance inducing transformation of pneumococcal types. Induction of transformation by a desoxyribonucleic acid fraction isolated from *Pneumococcus* Type III. *J. Exp.Med.* 79, 137–158 (1944).

Bush, G.W., "Stem Cell Science and the Preservation of Life, " *The New York Times*, August 12, 2001, p. D13.

Cohen, S. N., Chang, A.C., Boyer, H.W. & Helling, R. B.

Collins, F. & Galas, D. A new five-year plan for the US Human Genome Project. *Science* 262, 43–46 (1993).

Collins, F. S. *et al.*New goals for the US Human Genome Project: 1998–2003. *Science* 282, 682–689 (1998).

Drews, J.& Ryser, S. The role of innovation in drug development.

Druker, B. J. Imatinib alone and in combination for chronic myeloid leukemia. *Semin.Hematol.* 40, 50–8 (2003).

Fuller, B. P. *et al.* Policy forum: Ethics — privacy in genetics research. *Science* 285, 1359–1361 (1999).

Gabriel, S. B. *et al.* The structure of haplotype blocks in the human genome. *Science* 296, 2225–2229 (2002).

Gardner, M. J. *et al.* Genome sequence of the human malaria parasite *Plasmodium falciparum*. *Nature* 419, 498–511 (2002).

Gavin, A. C. *et al*. Functional organization of the yeast proteome by systematic analysis of protein complexes. *Nature* 415, 141–147 (2002).

Golub, T.R.Genomic approaches to the pathogenesis of hematologic malignancy.*Curr.Opin.Hematol.* 8, 252–261 (2001).

Guttmacher and Mark S.Guyer are at the National Human Genome Research Institute, National Institutes of Health, Bethesda, Maryland 20892, USA.

Guttmacher, A. E.& Collins, F. S. Genomic medicine — A primer.*N.Engl. J.Med.* 347, 1512–1520 (2002).

Hannon, G. J.RNA interference.*Nature* 418, 244–251 (2002).

Hilbert, D.Mathematical problems. *Bull.Am.Math. Soc.* 8, 437–479 (1902).

Hirschhorn, J.N., Lohmueller, K., Byrne, E. & Hirschhorn, K.A comprehensive review of genetic association studies. *Genet.*

Holt, R. A. *et al*. The genome sequence of the malaria mosquito *Anopheles gambiae*. *Science* 298, 129–149 (2002).

Horii, T., Nagao, Y., Tokunaga, T., and Imai, H. 20003. Serum-free culture of murine primordial germ cells and embryonic germ cells. Theriogen. 59, 1257-1264.

Horii, T., Nagao, Y., Tokunaga, T., and Imai, H. 20003. Serum-free culture of murine primordial germ cells and embryonic germ cells. Theriogen. 59, 1257-1264.

Jackson, D. A., Symons, R. H. & Berg, P. Biochemical method for inserting new genetic information into DNA of Simian Virus 40: circular SV40 DNA molecules containing lambda phage genes and the galactose operon of *Escherichia coli*. *Proc.Natl Acad. Sci.*

Kaczorowski, D.J., Patterson, E.S., Jastromb, W.E., Shamblott, M.J. 2002. Review Article. Glucose-responsive insulin-producing cells from stem cells. Diabetes Metab.

Kaczorowski, D.J., Patterson, E.S., Jastromb, W.E., Shamblott, M.J. 2002. Review Article. Glucose-responsive insulin-producing cells from stem cells. Diabetes Metab.

Kalka, C. et al. Transplantation of ex vivo expanded endothelial progenitor cells for therapeutic neovascularization. Proc. Natl. Acad. Sci. USA 2000; 97:3422-3427.

Kalkman EAJ, Bilgin YM, van Haren P, van Suylen R-J, Saxena PR, Schoemaker RG.Determinants of coronary reserve in rats subjected to coronary artery ligation or aortic banding. Cardiovasc Res 1996.

Kawano H, Do YS, Kawano Y, Starnes V, Barr M, Law RE, Hsueh WA. Angiotensin II has multiple profibrotic effects in human cardiac fibroblasts.

Kehat I, Kenyagin Karsenti D, Snir M, Segev H, Amit M, Gepstein A, Livne E, Binah O, Itskovitz-Eldor J, Gepstein L. Human embryonic stem cells can differentiate into myocytes with structural and functional properties of cardiomyocytes.

Kellerman S, Moore JA, Zierhut W, Zimmer HG, Campbell J, Gerdes AM. Nuclear DNA content and nucleation patterns in rat cardiac myocytes from different models of cardiac hypertrophy. *J Mol Cell Cardiol,* 497-505 (1992).

Kennedy M, Firpo M, Choi K, Wall C, Robertson S, Kabrun N, Keller G. A common precursor for primitive erythropoiesis and definitive haematopoiesis.

Kerr, D.A., Lladó, J., Shamblott, M.J., Maragakis, N.J., Irani, D.N., Crawford, T.O, Krishnan, C., Dike, S., Gearhart, J.D., and Rothstein, J.D. 2003. Human embryonic germ cell derivatives facilitate motor recovery of rats with diffuse motor neuron injury. J. Neurosci. 23, 5131-5140.

Kerr, D.A., Lladó, J., Shamblott, M.J., Maragakis, N.J., Irani, D.N., Crawford, T.O, Krishnan, C., Dike, S., Gearhart, J.D., and Rothstein, J.D. 2003. Human embryonic germ cell derivatives facilitate motor recovery of rats with diffuse motor neuron injury. J. Neurosci. 23, 5131-5140.

Kimura, T., Suzuki, A., Fujita, Y., Yomogida, K., Lomeli, H., Asada, N., Ikeuchi, M., Nagy, A., Mak, T.W., and Nakano, T. 2003. Development and Disease. Conditional loss of PTEN leads to testicular teratoma and enhances embryonic germ cell production. Development 130, 1691-1700.

Kimura, T., Suzuki, A., Fujita, Y., Yomogida, K., Lomeli, H., Asada, N., Ikeuchi, M., Nagy, A., Mak, T.W., and Nakano, T. 2003. Development and Disease. Conditional loss of PTEN leads to testicular teratoma and enhances embryonic germ cell production. Development 130, 1691-1700.

Klug MG, Soonpaa MH, Koh GY, Field LJ (1996) Genetically selected cardiomyocytes from differentiating embryonic stem cells form stable intracardiac grafts. J Clin Invest 98:216-224.

Kocher A, Schuster M, Szabolcs M, Itescu S. Cardiomyocyte regeneration after neovascularization of ischemic myocardium by human bone marrowderived angioblasts. Nature Medicine 2002.

Kocher AA, Schuster MD, Szabolcs MJ, Takuma S, Burkhoff D, Wang J, Homma S, Edwards NM, Itescu S. Neovascularization of ischemic myocardium by human bone-marrow-derived angioblasts prevents cardiomyocyte apoptosis, reduces remodeling and improves cardiac function. Nature Med.

L Field: Future therapy for cardiovascular disease. In Proceedings of the NHLBI Workshop Cell Transplantation: Future Therapy for Cardiovascular Disease? Columbia, MD; 1998.

Labastie M-C, Cortes F, Romeo P-H, Dulac C, Peault B. Molecular identity of hematopoietic precursor cells emerging in the human embryo. Blood 1998;92:3624-3635.

Levkau B, Koyama H, Raines EW, Clurman BE, Herren B, Orth K, Roberts JM, Ross R. Cleavage of p21cip1/waf1 and p27 kip1 mediates apoptosis in endothelial cells through activation of cdk2: role of a caspase cascade.

Liechty KW, MacKenzie TC, Shaaban AF, et al. Human mesenchymal stem cells engraft and demonstrate site-specific differentiation after in utero transplantation in sheep. Nat Med 2000, 6:1282-6.

Lin Y, Weisdorf DJ, Solovey A, Hebbel RP. Origins of circulating endothelial cells and endothelial outgrowth from blood.

Lynch, H. T. & de la Chapelle, A. Genomic medicine: hereditary colorectal cancer *N.Engl. J.Med.* 348, 919–932 (2003).

Lysaght, M.J., and Hazlehurst, A.L., "Private Sector Development of Stem Cell Technology and Therapeutic Cloning, " *Tissue Engineering* 9(3): 555-561 (2003).

MacLellan WR and Schneider MD. Genetic dissection of

cardiac growth control pathways Annu. Rev. Physiol. 2000. 62:289-320.

Mahon NG, O'Roke C, Codd MB, et al. Hospital mortality of acute myocardial infarction in the thrombolytic era. Heart 1999;81:478-82.

Maier S, Tertilt C, Chambron N, Gerauer K, Huser N, Heidecke C-D, Pfeffer K (2001) Inhibition of natural killer cells results in acceptance of cardiac allografts in CD28-/- mice. Nature Med 7:557-562.

Makino S, Fukuda K, Miyoshi S, et al. Cardiomyocytes can be generated from marrow stromal cells in vitro. J Clin Invest 1999, 103:697-705.

Maxam, A. M. & Gilbert, W. A new method for sequencing DNA.

McCarthy PM, Rose EA, Macmanus Q, et al. Clinical experience with the Novacor ventricular assist system. J Thorac Cardiovasc Surg. 1991; 102:578-587.

McEwan PE, Gray GA, Sherry L, Webb DJ, Kenyon CJ. Differential effects of angiotensin II on cardiac cell proliferation and intramyocardial perivascular fibrosis in vivo. Circulation 1998;98:2765-2773.

Mendel, G.Versuche über Pflanzen-Hybriden.*Verhandlungen des naturforschenden Vereines, Abhandlungen, Brünn* 4, 3–47 (1866).

Migeon, B.R., Chowdry, A.K., Dunston, J.A., and McIntosh, I. 2001. Identification of *TSIX*, encoding an RNA antisense to human *XIST*, reveals differences from its murine counterpart: Implications for X inactivation. Am. J. Human Genet. 69, 951-960.

Migeon, B.R., Chowdry, A.K., Dunston, J.A., and McIntosh, I. 2001. Identification of *TSIX*, encoding an RNA antisense to human *XIST*, reveals differences from its murine counterpart: Implications for X inactivation. Am. J. Human Genet. 69, 951-960.

Milki, A.A. et al., "Incidence of monozygotic twinning with blastocyst transfer compared to cleavage-stage transfer" *Fertility and Sterility* 79(3): 503-506 (2003).

Miller, P. S. Is there a pink slip in my genes? *J.Health Care Law Policy* 3, 225–265 (2000).

National Center for Health Statistics, "Births: Final Data for 2001." National Vital Statistics Reports 51(2) (2002), available at http://www.cdc.gov/nchs/data/nvsr/ nvsr51/ nvsr51_02.pdf.

National Institutes of Health Revitalization Act of 1993, Pub. L. No. 103-43, § 121(c), 107 Stat. 122 (1993), repealing 45 C.F.R. § 46.204(d).

National Institutes of Health, *Report of the Human Embryo Research Panel*, Bethesda, MD: NIH (1994).

National Research Council.*Mapping and Sequencing the Human Genome* (National Academy Press, Washington DC, 1988).

Nelissen-Vrancken H, Debets J, Snoeckx L, Daemen M, Smits J. Time-related normalization of maximal coronary flow in isolated perfused hearts of rats with myocardial infarction. Circulation 1996;93:349-355.

Nidorf SM, Siu SC, Galambos G, Weyman AE, Picard MH. Benefit of late coronary reperfusion on ventricular morphology and function after myocardial infarction. J Am Coll Cardiol 1992;20:307-313.

Nirenberg, M.W. The genetic code: II. *Sci.Am.* 208, 80–94 (1963).

Norwitz, E. R., et al., "Implantation and the survival of early pregnancy, " *New England Journal of Medicine* 345(19): 1400-1408 (2001)

Ogawa M, Kizumoto M, Nishikawa S, Fujimoto T, Kodama H, Nishikawa SI. Blood 1999;93:1168-1177.

Onyango, P., Jiang, S., Uejima, H., Shamblott, M.J., Gearhart, J.D., Cui, H., and Feinberg, A.P. 2002. Monoallelic expression and methylation of imprinted genes in human and mouse embryonic germ cell lineages. Proc. Natl. Acad. Sci. USA 99, 10599-10604.

Onyango, P., Jiang, S., Uejima, H., Shamblott, M.J., Gearhart, J.D., Cui, H., and Feinberg, A.P. 2002. Monoallelic expression and methylation of imprinted genes in human and mouse embryonic germ cell lineages. Proc. Natl. Acad. Sci. USA 99, 10599-10604.

Oz MC, Argenziano M, Catanese KA, et al. Bridge experience with long-term implantable left ventricular assist devices. Are they an alternative to transplantation? Circulation. 1997; 95:1844-1852.

Park, T.S., Hong, Y.H., Kwon, S.C., Lim, J.M., and Han, J.Y. 2003. Birth of germline chimeras by transfer of chicken embryonic germ (EG) cells into recipient embryos.

Park, T.S., Hong, Y.H., Kwon, S.C., Lim, J.M., and Han, J.Y. 2003. Birth of germline chimeras by transfer of chicken embryonic germ (EG) cells into recipient embryos.

Pfeffer JM, Pfeffer MA, Fletcher PJ, Braunwald E. Progressive ventricular remodeling in rat with myocardial infarction. Am J Physiol 1991.

Pfeffer MA, Braunwald E, Moye LA, et al. Effect of captopril on mortality and morbidity in patients with left ventricular dysfunction after myocardial infarction. Results of the survival and ventricular enlargement trial. The SAVE investigators. N Engl J Med 1992;327:669-677.

Pitt B, Segal R, Martinez FA, et al. Randomised trial of

losartan versus captopril in patients over 65 with heart failure (Evaluation of Losartan in the Elderly Study, ELITE). Lancet 1997;349:747-752.

Pittenger MF, Mackay AM, Beck SC, et al. Multilineage potential of adult human mesenchymal stem cells. Science 1999, 284:143-147.

Pouzet B, Ghostine S, Vilquin J-T, Garcin I, Scorsin M, Hagege AA, Duboc D, Schwartz K, Menasche P (2001) Is skeletal myoblast transplantation clinically relevant in the era of angiotensin-converting enzyme inhibitors? Circulation 104:I223-228.

Rafii S, Shapiro F, Rimarachin J, Nachman R, Ferris B, Weksler B, Moore AS, Asch AS. Isolation and characterization of human bone marrow microvascular endothelial cells: hematopoietic progenitor cell adhesion.

Ravichandran LV and Puvanakrishnan R. In vivo labeling studies on the biosynthesis and degradation of collagen in experimental myocardial myocardial infarction. Biochem Intl 1991;24:405-414.

Reich, D. E. & Lander, E. S. On the allelic spectrum of human disease. *Trends Genet.* 17, 502–510 (2001).

Rothenberg, K. *et al.* Genetic information and the workplace: Legislative approaches and policy challenges. *Science* 275, 1755–1757 (1997).

Sachidanandam, R. *et al.* A map of human genome sequence variation containing 1.42 million single nucleotide polymorphisms. *Nature* 409, 928–933 (2001).

Sanger, F.& Coulson, A. R. A rapid method for determining sequences in DNA by primed synthesis with DNA polymerase.

Sapienza, C. 2002. Commentary. Imprinted gene expression, transplantation medicine, and the "other" human embryonic stem cell. Proc. Natl. Acad. Sci. USA 99, 10243-10245.

Sapienza, C. 2002. Commentary. Imprinted gene expression, transplantation medicine, and the "other" human embryonic stem cell. Proc. Natl. Acad. Sci. USA 99, 10243-10245.

Selkoe, D. J.Alzheimer's disease: genes, proteins, and therapy. *Physiol. Rev.* 81, 741–66 (2001).

Shamblott, M.J., et al., "Derivation of pluripotent stem cells from cultured human primordial germ cells" *Proceedings of the National Academy of Sciences*. 95(23): 13726-13731 (1998).

Shi Q, Rafii S, Wu MH-D, et al. Evidence for circulating bone marrow-derived endothelial cells. Blood 1998;92:362-367.

Sidow, A. Sequence first.Ask questions later. *Cell* 111, 13–16 (2002).

Smith, L. M. *et al.* Fluorescence detection in automated DNAsequence analysis. *Nature* 321, 674–679 (1986).

Smythe GM, Grounds MD (2000) Exposure to tissue culture conditions can adversely affect myoblast behaviour in vivo in whole muscle grafts: implications for myoblast transfer therapy. Cell Transplant 9: 379-393.

Smythe G.M., Hodgetts S.I., Grounds M.D. (2001) Problems and solutions in myoblast transfer therapy. J Cell Mol Med 5:33-47.

Stockwell, B. R. Chemical genetics: Ligand-based discovery of gene function. *Nature Rev. Genet.* 1, 116–125 (2000).

Takahashi, T. et al. Ischemia- and cytokine-induced mobilization of bone marrowderived endothelial progenitor cells for neovascularization. Nat. Med. 1999; 5:434-438.

Tyson, J.J., Chen, K. & Novak, B.Network dynamics and cell physiology.*Nature Rev.Mol. Cell Biol.* 2, 908–916 (2001).

Van der Weyden, L., Adams, D. J. & Bradley, A. Tools for targeted manipulation of the mouse genome. *Physiol.Genomics* 11, 133–164 (2002).

Wagner, K.R. Genetic diseases of muscle.*Neurol. Clin.* 20, 645–678 (2002).

Watson, J.D. & Crick, F.H. C. Molecular structure of nucleic acids: A structure for deoxyribose nucleic acid. *Nature* 171, 737 (1953).

Wilcox, A.J., et al., "Incidence of early loss of pregnancy, " *New England Journal of Medicine* 319 (4): 189-194 (1988).

Xu, R.H., et al., "BMP4 initiates human embryonic stem cell differentiation to trophoblast" *Nature Biotechnology* 20(12): 1261-1264 (2002).

Index